# Wireless Networks

**Series editor**

Xuemin Sherman Shen
*University of Waterloo, Waterloo, Ontario, Canada*

The purpose of Springer's new Wireless Networks book series is to establish the state of the art and set the course for future research and development in wireless communication networks. The scope of this series includes not only all aspects of wireless networks (including cellular networks, WiFi, sensor networks, and vehicular networks), but related areas such as cloud computing and big data. The series serves as a central source of references for wireless networks research and development. It aims to publish thorough and cohesive overviews on specific topics in wireless networks, as well as works that are larger in scope than survey articles and that contain more detailed background information. The series also provides coverage of advanced and timely topics worthy of monographs, contributed volumes, textbooks and handbooks.

More information about this series at http://www.springer.com/series/14180

Hongliang Zhang • Lingyang Song • Zhu Han

# Unmanned Aerial Vehicle Applications over Cellular Networks for 5G and Beyond

Springer

Hongliang Zhang
School of Electrical Engineering
and Computer Science
Peking University
Beijing, China

Lingyang Song
School of Electrical Engineering
and Computer Science
Peking University
Beijing, China

Zhu Han
Department of Electrical and Computer
Engineering
University of Houston
Houston, TX, USA

ISSN 2366-1186 ISSN 2366-1445 (electronic)
Wireless Networks
ISBN 978-3-030-33038-5 ISBN 978-3-030-33039-2 (eBook)
https://doi.org/10.1007/978-3-030-33039-2



This Springer imprint is published by the registered company Springer Nature Switzerland AG.
The registered company address is: Gewerbestrasse 11, 6330 Cham, Switzerland

# Preface

The emerging unmanned aerial vehicles (UAVs) are playing an increasing role in military, public, and civil applications. Very recently, 3GPP has approved a study item to facilitate seamless integration of UAVs into future cellular networks. Unlike terrestrial cellular networks, UAV communications have many distinctive features such as high dynamic network topologies and weakly connected communication links. In addition, they still suffer from some practical constraints such as battery power and no-fly zone. As such, many standards, protocols, and design methodologies used in terrestrial wireless networks are not directly applicable to airborne communication networks. Therefore, it is essential to develop new communication, signal processing, and optimization techniques in support of the ultra-reliable and real-time sensing applications, but enabling high data rate transmissions to assist the terrestrial communications in the future cellular networks. Typically, to integrate UAVs into cellular networks, one needs to consider two main scenarios of UAV applications as follows.

First, dedicated UAVs can be used as communication platforms in the way as wireless access points or relays nodes to further assist the terrestrial communications. This type of applications can be referred to as *UAV Assisted Cellular Communications*. UAV assisted cellular communications have numerous use cases, including traffic offloading, wireless backhauling, swift service recovery after natural disasters, emergency response, rescue and search, information dissemination/broadcasting, and data collection from ground sensors for machine-type communications. However, different from traditional cellular networks, how to plan the time-variant placements of the UAVs served as base station (BS)/relay is very challenging due to the complicated 3D propagation environments as well as many other practical constraints such as power and flying speed. In addition, spectrum sharing with existing cellular networks is another interesting topic to investigate.

The second type of application is to exploit UAVs for sensing purposes due to its advantages of on-demand flexible deployment, larger service coverage compared with the conventional fixed sensor nodes, and the ability to hover. Specially, UAVs, equipped with cameras or sensors, have come into our daily lives to execute critical real-time sensing tasks, such as smart agriculture, security monitoring, forest fire detection, and traffic surveillance. Due to the limited computation capability of UAVs, the sensory data needs to be transmitted to the BS for real-time data processing. In this regard, the cellular networks are necessarily committed to supporting the data transmission for UAVs, which we refer to as *Cellular assisted UAV Sensing*. Nevertheless, to support real-time sensing streaming, it is desirable to design joint sensing and communication protocols, develop novel beamforming and estimation algorithms, and study efficient distributed resource optimization methods.

The aim of this book is to educate control, signal processing engineers, computer and information scientists, applied mathematicians and statisticians, as well as systems engineers to carve out the role that analytical and experimental engineering has to play in UAV research and development. This book will emphasize on UAV technologies and applications for cellular networks.

To summarize, the key features of this book are as follows:

1. It provides an introduction to the UAV paradigm, from 5G and beyond communication perspective, which currently has attracted plenty of attention from both academic and industrial sides.
2. It introduces the key methods, including optimization, game, and machine learning, for UAV applications, in a comprehensive way. It also discusses the state-of-the-art for cellular network assisted UAV sensing. Many examples will be illustrated in detail so as to provide a wide scope for general readers.
3. It includes formalized analysis of several up-to-date networking challenges. Some machine learning methods are introduced to effectively solve these challenges. These successful cases show how the machine learning can benefit and accelerate the UAV network development.

This book is organized as below. Chapter 1 provides an overview of the UAV paradigm and discusses the state-of-the-art. In Chap. 2, we review some key methods which will be used for UAV applications, including optimization, game, and machine learning. In Chaps. 3 and 4, we give the study cases to show how to solve the key challenges in UAV assisted cellular communications and cellular assisted UAV sensing, respectively.

Beijing, China Hongliang Zhang
Beijing, China Lingyang Song
Houston, TX, USA Zhu Han

# Contents

# Acronyms

| | |
|---|---|
| 2D | Two dimensional |
| 3D | Three dimensional |
| 4G | Fourth generation |
| 5G | Fifth generation |
| AQI | Air quality index |
| AWGN | Additive white Gaussian noise |
| AP | Access point |
| AF | Amplify-and-forward |
| ATC | Air traffic control |
| A2G | Air-to-ground |
| A2A | Air-to-air |
| AEA | Average estimation accuracy |
| BS | Base station |
| CU | Cellular user |
| CNPC | Control and non-payload communications |
| CDF | Cumulative distribution function |
| D2D | Device-to-device |
| DF | Decode-and-forward |
| DC | Difference of convex functions |
| DQN | Deep Q-network |
| EE | Energy efficiency |
| eMBB | Enhanced mobile broadband |
| F-Cell | Flying-cell |
| GSC | Ground control station |
| GPI | Generalized policy iteration |
| GPM-NN | Gaussian plume model embedding neural networks |
| IoT | Internet-of-things |
| ITU-R | International telecommunication union—Radiocommunications standardization sector |
| IC | Incentive compatibility |
| IR | Individual rationality |

| | |
|---|---|
| IP | Increasing preference |
| KKT | Karush–Kuhn–Tucker |
| LoS | Line-of-sight |
| LTE | Long-term evolution |
| MBS | Macro-cell base station |
| MD | Mobile device |
| mMTC | Massive machine-type communications |
| MDP | Markov decision processes |
| NLoS | Non-line-of-sight |
| NN | Neural networks |
| PPP | Poisson point process |
| PoI | Point of Interests |
| PDF | Probability density function |
| PC | Payload communication |
| PDT | Partial derivative threshold |
| QoS | Quality of services |
| RMa | Rural macro |
| SBS | Small-cell base station |
| SNR | Signal-to-noise ratio |
| SINR | Signal-to-interference-plus-noise ratio |
| SVM | Support vector machine |
| TD | Temporal difference |
| TSP | Traveling salesman problem |
| U2X | UAV-to-everything |
| U2N | UAV-to-network |
| U2U | UAV-to-UAV |
| UAV | Unmanned aerial vehicle |
| URLLC | Ultra-reliable and low-latency communications |
| UAS | Unmanned aircraft systems |
| UMa | Urban macro |
| UMi | Urban micro |
| UE | User equipment |

# Chapter 1
# Overview of 5G and Beyond Communications

Unmanned aerial vehicles (UAVs) are aircrafts piloted by remote control or embedded programs without human onboard. During the 1930s, the US Navy began to experiment with radio-controlled UAVs. From the 1990s, micro UAVs started to be widely used in public and civilian applications. Recently, due to the ease of deployment, low acquisition and maintenance costs, high maneuverability, and ability to hover, UAVs have been widely used in civil and commercial applications [1]. However, traditional UAV research has typically focused on navigation [2] and autonomy [3], as it is motivated by the military oriented applications. In contrast, this book will focus on sensing, communication, and learning for UAV applications over future cellular networks.

In this chapter, we will first introduce the background and requirements in Sect. 1.1, and then different UAV applications over cellular networks are presented in Sect. 1.2. Finally, we will elaborate on current state-of-the-art in Sect. 1.3.

## 1.1 Background and Requirements

Following the global commercial success of fourth generation (4G) mobile communications based on the Long Term Evolution (LTE)-Advanced standard developed by the Third Generation Partnership Project (3GPP), the industry together with the research community have been increasingly exploring fifth generation (5G) mobile communications. International Telecommunication Union—Radiocommunications Standardization Sector (ITU-R) has developed a vision for 5G mobile communications that includes support for enhanced mobile broadband (eMBB) to accommodate explosive growth of data traffic, support for massive machine-type communications (mMTC), and support for ultra-reliable and low-latency communications (URLLC) [4].



H. Zhang et al., *Unmanned Aerial Vehicle Applications over Cellular Networks for 5G and Beyond*, Wireless Networks, https://doi.org/10.1007/978-3-030-33039-2_1

To support the 10× peak data rate requirement of 5G eMBB, according to the Shannon formula, we have the following solutions:

- Increase the number of access points (APs).
- Increase the bandwidth.
- Increase the signal-to-interference-plus-noise ratio (SINR).

In this regard, UAV assisted cellular communications are one of the promising solutions for the data rate enhancement. First, UAVs can serve as flying APs to deliver reliable and cost-effective wireless communications, and thus densifying the APs. Second, the UAVs can also serve as wireless relays to improve the SINR, especially for the cell-edge users.

Another typical feature of future cellular networks is the integrations of massive connected Internet-of-Things (IoT) devices, which connects physical world by sensors. The analyst firm Gartner estimated that over 2020 billion connected things are expected to be in use by 2020 [5]. Recently, UAVs equipped with sensors have come into our daily life to execute a variety of sensing missions, due to their ease of deployment, high autonomy, and the ability to hover [6]. To support the reliable UAV sensing applications, the terrestrial cellular networks are promising enabler, which we referred to as cellular assisted UAV sensing. Motivated by these reasons, it is necessary to integrate UAV applications to cellular networks.

## 1.2 UAV Applications

Generally, the UAVs can be classified into two types: fixed-wings and rotary-wings.

- Rotary-wing UAVs allow vertical take-off and landing, and can hover over a fixed location. This high maneuverability makes them suitable to execute sensing tasks. Also, the rotary-wing UAVs can be used to deploy base station (BS) since they can hover at the desired locations with higher precision. However, rotary-wings UAVs have limited mobility and power.
- Fixed-wing UAVs can glide over the air, which makes them significantly more energy efficient and able to carry heavy payload. Therefore, the fix-wing UAVs can carry cellular infrastructure to provide cellular coverage. Gliding also helps fixed-wing UAVs to fly at a faster speed, which makes them suitable to be applied to airborne surveillance. The disadvantage of fixed-wing UAVs is that they cannot hover over a fixed location. Fixed-wing UAVs are also more expensive than rotary-wing one [7].

The characteristics of the UAVs are given in Table 1.1. According to the characteristics of the UAVs, UAVs have two kinds of applications: flying infrastructure and aerial IoT.

**Table 1.1** Characteristics of different UAVs

| | Fixed-wing UAVs | Rotary-wing UAVs |
|---|---|---|
| Speed | Up to 500 km/h | Typically less than 60 km/h |
| Altitude | Up to 20 km | Typically less than 1 km |
| Flight time | Up to several hours | Typically less than 30 min |
| Applications | Carry cellular infrastructure | Sensing |
| | Airborne surveillance | Carry cellular BS |

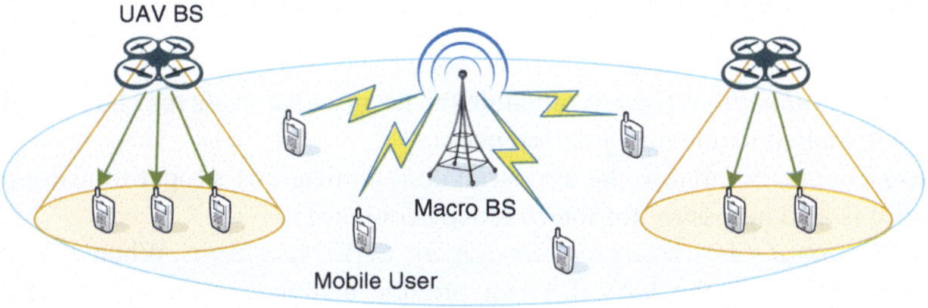


**Fig. 1.1** UAV serving as BS

## *1.2.1 Flying Infrastructure*

Due to the low operational cost and flexible deployment, UAVs are adopted as flying infrastructure to provide wireless communications. For example, Nokia's flying-cell (F-Cell) [8] and Facebook's Aquila [9] are two existing projects using UAVs to provide the wireless coverage. In the Facebook's Aquila project, UAVs with solar panels are sent 18–27 km into the stratosphere to provide widespread Internet coverage, which can reach 40–80 km$^2$.

In these applications, there exist two main scenarios: UAV as flying BS and relay.

**UAV as BS** As shown in Fig. 1.1, the UAV can work as the flying BS in hotspot for capacity enhancement or data offloading [20]. Also, the UAV can be used as a temporary BS for emergency communications due to its high mobility and the ease of deployment.

**UAV as Relay** In Fig. 1.2, we present a system where a UAV serves as the relay to extend coverage or aggregate data [21]. The UAV can utilize amplify-and-forward (AF) or decode-and-forward (DF) protocol to relay the data. Different from UAV BS, where the backhauling is assumed to be perfect, the backhauling link for UAV relaying should also be considered.

In the application, the objective is to guarantee the QoS of UAV communications, and thus the following research challenges should be well addressed:

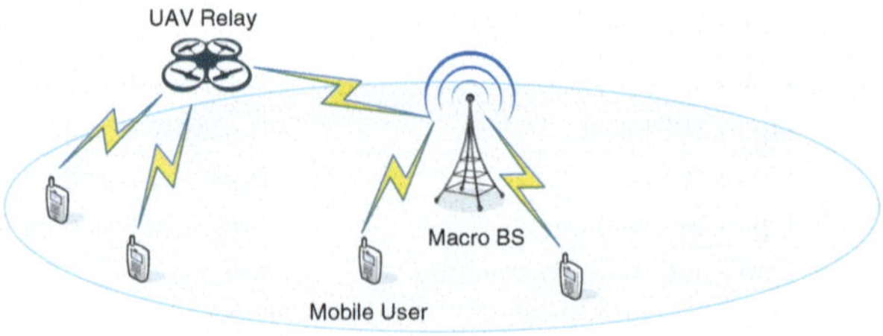


**Fig. 1.2** UAV serving as relay

1. *Frequency allocation*: How to schedule the users and allocate the channels should be well studied to alleviate interference.
2. *Power control*: Similar to the terrestrial communications, proper transmit power control is also necessary for interference avoidance.
3. *3-Dimensional (3D) trajectory design or UAV placement*: When the UAV's altitude is higher, the UAV BS can provide a higher coverage, but the data rate will reduce because it will suffer a higher pathloss. Therefore, the trajectory/placement of UAV should be well designed to achieve the trade-off between the coverage and the data rate.

## *1.2.2 Aerial Internet-of-Things*

Unmanned Aircraft Systems (UAS) is an existing technology to enable the various civilian applications. The UAS consists of the UAVs, typically rotary-wing UAVs, and ground control station (GSC). The UAVs are equipped with payload, such as camera, thermometer, and Air Quality Index (AQI) sensor, to execute real-time sensing tasks, and transmit the sensory data to the GSC or server. To support the efficient operations of the UAS, there are two kinds of transmissions: payload communication (PC) and control and non-payload communication (CNPC).

- *CNPC*: CNPC refers to the communications between UAVs and GSCs for the reliable and effective operations. Typical CNPC messages include [10]:

  1. Remote command and control for UAVs;
  2. Navigation aids;
  3. Surveillance data for sense-and-avoid and weather radar data;
  4. Air traffic control (ATC) information relaying.

  CNPC is usually of low data rate requirement, e.g., 7 kbps for an uplink medium/large UAV, but has a strict requirement on high reliability, high security, and low latency.

- *PC*: PC refers to sensory data transmission between UAVs and GSCs, such as real-time video and image. Compared to CNPC, PC usually has much higher data rate requirements. For example, to transmit a full-high-definition (FHD) video, the data rate should be up to several Mbps.

The existing UAS systems are operated over the unlicensed spectrum for CNPC and PC. For example, the working bands for DJI Phantom 4 Pro are 2.4–2.483 GHz and 5.725–5.825 GHz, which are in the Industrial, Scientific and Medical (ISM) Bands. However, the quality-of-services (QoS) of communications is not guaranteed in the ISM bands because the interference from the environment is not controllable. Also, the communication range is limited.

Therefore, terrestrial cellular network is a promising solution to support UAV sensing applications due to the larger coverage, higher reliability and flexibility, which we refer to as the cellular Internet of UAVs [11]. In the cellular Internet of UAVs, the UAVs transmit the sensory data to the BS by two communication modes: UAV-to-Network (U2N) communications, where UAVs transmit the sensory data to the BS directly, and UAV-to-UAV (U2U) communications, where UAVs transmit to the BS through a UAV relay, as shown in Fig. 1.3.

With the objective to guarantee the QoS for both sensing and communication, there exist the following research challenges in the applications as aerial IoT components:

- *Trajectory optimization*: Since the trajectory of a UAV will influence the QoS of sensing and communications, 3D trajectory optimization is necessary.
- *Frequency allocation*: To avoid the interference among different UAVs, proper frequency allocation or UAV transmission scheduling schemes should be well designed.
- *Power control*: The energy is also a critical problem for the UAVs because the on-board battery is limited. Therefore, how to balance the transmission energy and propulsion energy should be well investigated.

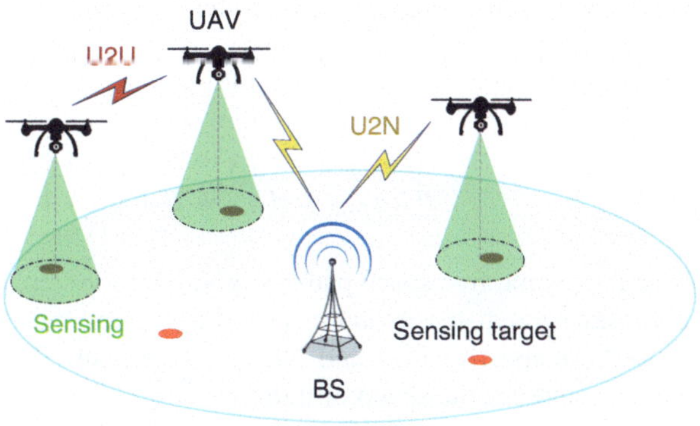


**Fig. 1.3** Cellular internet of UAVs

## 1.3 Current State-of-the-art

In this section, we first introduce the channel model for UAV communications over cellular networks, and then analyze the network performance. Finally, we present the propulsion and mobility model.

### *1.3.1 Channel Model*

According to the transmission modes, the channel model can be categorized into two types, i.e., Air-to-Ground (A2G) and Air-to-Air (A2A) channels. Since the communications between UAVs typically occur in clear airspace, and thus the A2A channel can be characterized by the free space pathloss model [12]. Therefore, in this subsection, we focus on the A2G channels, where UAVs communicate with ground mobile devices or BSs.

The A2G channels significantly differ from terrestrial communication channels. Any movement caused by the UAVs can influence the channel characteristics. To be specific, the channel characteristics are highly related to the altitudes and elevation angles of the UAVs. A larger elevation angle will lead to a lower pathloss with the same propagation distance. In addition, the channels are also affected by the propagation environment. Light-of-Sight (LoS) links are expected in most scenarios, while they can also be occasionally blocked by obstacles such as terrain, buildings, or the airframe itself. To capture these features, probability pathloss models have been widely adopted for the A2G channel, in which the LoS and Non-Light-of-Sight (NLoS) links are considered with different probabilities of occurrence. In the following, we will present two well-known models, i.e., *elevation angle-based model* [13] and *3GPP model* [14].

#### 1.3.1.1 Elevation Angle-Based Model

In this model, the LoS and NLoS pathloss in dB is modeled by

$$\begin{aligned} PL_{LoS} &= L_{FS} + 20\log(d) + \eta_{LoS}, \\ PL_{NLoS} &= L_{FS} + 20\log(d) + \eta_{NLoS}, \end{aligned} \tag{1.1}$$

where $L_{FS}$ is the free space pathloss given by $L_{FS} = 20\log(f) + 20\log(\frac{4\pi}{c})$, $d$ is the transmission distance, and $\eta_{LoS}$ and $\eta_{NLoS}$ are the additional attenuation factors due to the respective LoS and NLoS components. Here, $f$ is the system carrier frequency and $c$ is the speed of light.

The probability of the LoS component depends on the environment and elevation angle $\theta$ between the UAV and ground device, which is modeled as a logistic function of $\theta$:

$$P_{LoS} = \frac{1}{1 + a\exp(-b(\theta - a))}, \tag{1.2}$$

where $a$ and $b$ are constants which depend on the environment. The probability of the NLoS component is given by

$$P_{NLoS} = 1 - P_{LoS}. \tag{1.3}$$

With such a model, the average pathloss can be expressed by

$$PL_{avg} = P_{LoS}PL_{Los} + P_{NLoS}PL_{NLoS}. \tag{1.4}$$

Moreover, the small-scale fading in the A2A channel can be characterized by the Rician fading model [15], which consists of a deterministic LoS component and a random scattered component.

#### 1.3.1.2 3GPP Model

The 3GPP channel model is provided for the communications between ground BS and the aerial users with the altitude varying from 1.5 to 300 m. In this model, three typical 3GPP scenarios are studied, namely Rural Macro (RMa), Urban Macro (UMa), and Urban Micro (UMi).

Different from the elevation angle-based model which only considers the effect of the elevation angle, the 3GPP model is the combination of both the elevation angle and the altitude. For these three scenarios, the probability of the LoS component can be specified by the 2D distance between the ground BS and the UAV $d_{2D}$, as well as the altitude of the UAV $H_{UT}$. If $H_{UT}$ is lower than a predefined threshold $H_1$, the aerial user can be regarded as a terrestrial user, and thus the channel models for terrestrial communications can be used directly. On the contrast, if $H_{UT}$ is higher than another threshold $H_2$, the probability of the LoS component can be 100%, and the probability of the LoS component will be a function $P_{LoS}(d_{2D}, H_{UT})$ related with $d_{2D}$ and $H_{UT}$ when $H_1 \leq H_{UT} \leq H_2$.

Mathematically, the probability of the LoS component $P_{LoS}$ can be written by

$$P_{LoS} = \begin{cases} P_{ter}, & 1.5m \leq H_{UT} < H_1, \\ P_{LoS}(d_{2D}, H_{UT}), & H_1 \leq H_{UT} \leq H_2, \\ 1, & H_2 < H_{UT} \leq 300m, \end{cases} \tag{1.5}$$

where $P_{ter}$ is the probability of the LoS component in terrestrial communications as given in Table 7.4.2 of [16], and $P_{LoS}(d_{2D}, H_{UT})$ is given by

$$P_{LoS}(d_{2D}, H_{UT}) = \begin{cases} 1, & d_{2D} \leq d_1, \\ \frac{d_1}{d_{2D}} + \exp(\frac{-d_{2D}}{p_1})(1 - \frac{d_1}{d_{2D}}), & d_{2D} > d_1, \end{cases} \tag{1.6}$$

with $p_1$ and $d_1$ being given constants corresponding to the scenarios as shown in Table B-1 of [14]. It is worthwhile to mention that $H_1$ and $H_2$ will be different in the aforementioned three scenarios. For example, $H_1 = 10$ m in the RMA model while $H_1 = 22.5$ m in the UMa model.

Moreover, based on the environment, the pathloss models for the LoS and NLoS components as well as the shadowing fading standard deviation are given in Table B-2 and Table B-3 of [14], respectively.

In Fig. 1.4, we give the average channel power versus UAV altitude $H_{UT}$ for different values of $d_{2D}$ in the RMa scenario, and the simulation parameters are given in Table 1.2. As shown in Fig. 1.4, the average channel power first increases with UAV altitude $H_{UT}$ because the probability of LoS component increases, and then it deceases when $H_{UT}$ exceeds a certain threshold, where the benefit of the increased LoS probability cannot make up for the increased pathloss due to a longer transmission distance.

In the following two subsections, we give some simulation results to help the readers to understand the characteristics of UAV communications. In these simulations, we use the aforementioned 3GPP channel model.

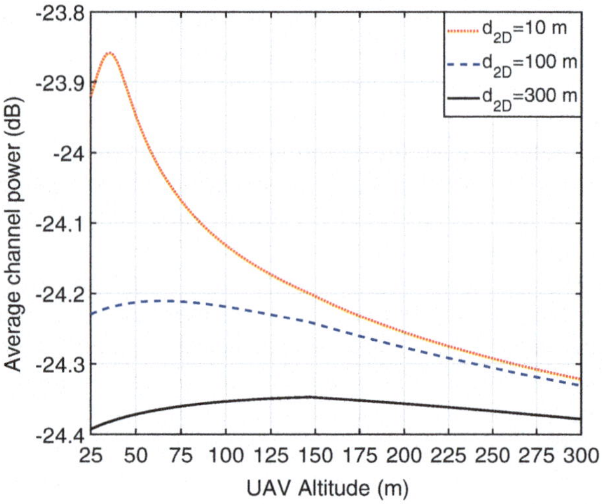


**Fig. 1.4** Average channel power versus UAV altitude in the RMa scenario of the 3GPP model

**Table 1.2** Parameters for channel model

| Parameters | Values |
|---|---|
| Height of BS | 25 m |
| Carrier frequency | 3 GHz |

### 1.3.2 Aerial Access Networks

In this subsection, we will present the performance analysis on aerial access networks where the UAVs are flying infrastructure.

1. **UAV relay**: As shown in Fig. 1.5, we consider the uplink of a single cell with one terrestrial user equipment (UE). To improve the performance of the UE, we introduce one UAV as an AF relay between the UE and the BS. The simulation parameters are given in Table 1.3.

   It is assumed that the UAV is static during the transmission process. It is obvious that the optimal location of the UAV relay is in the vertical plane determined by the UE and the BS. Therefore, we can assume that the UAV relay is located in that plane. In addition, we use $h_{relay}$ to represent the altitude of the UAV relay, and use $d_{UB}$ to denote the horizontal distance between the UAV relay and the BS (Table 1.3).

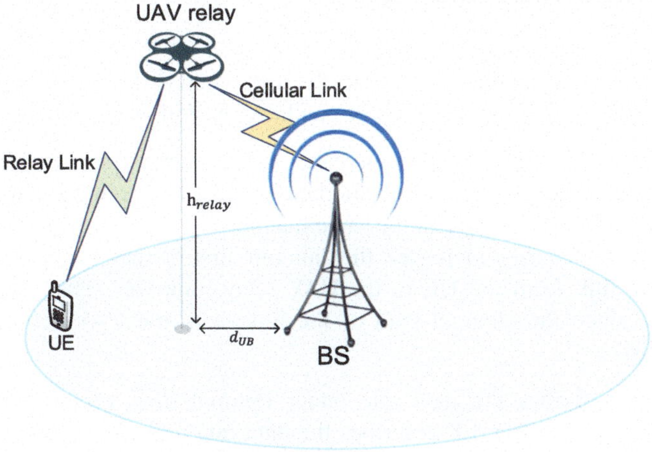


**Fig. 1.5** System model of the UAV relay network

**Table 1.3** Simulation parameters for UAV relay

| Parameters | Values |
|---|---|
| Height of BS | 25 m |
| Height of terrestrial UE | 1 m |
| Carrier frequency | 3 GHz |
| Transmit power of the UAV relay | 23 dBm |
| Transmit power of the UE | 23 dBm |
| Noise variance | −96 dBm |
| Antenna number of UE | 1 Tx, 1 Rx |
| Antenna number of UAV relay | 1 Tx, 1 Rx |
| Antenna number of BS | 1 Tx, 1 Rx |
| Antenna pattern of UE, UAV relay, and BS | Omnidirection |

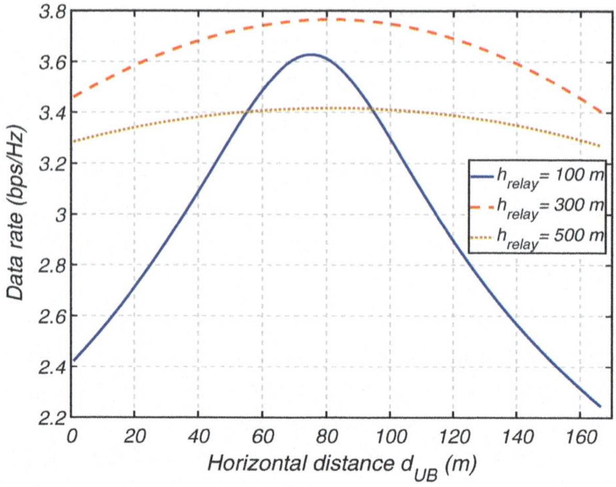


**Fig. 1.6** Data rate vs. horizontal distance $d_{UB}$

Figure 1.6 depicts the data rate versus the horizontal distance between the UAV relay and the BS $d_{UB}$. From Fig. 1.6, we can have the following observations:

1. As the altitude of the UAV relay increases, the data rate first increases and then decreases, since the channel gain from the UE to the UAV relay and that from the UAV relay to the BS change in the same manner.
2. When the distance $d_{UB}$ increases, the data rate first increases, since the channel gain of the link from the UE to the UAV relay improves. Then, the data rate decreases, since the channel gain of the link from the UAV relay to the BS deteriorates.

Figure 1.7 describes the data rate under optimal $d_{UB}$ versus height of the UAV relay $h_{relay}$, and Fig. 1.8 describes the data rate under optimal $h_{relay}$ versus horizontal distance $d_{UB}$. We can observe that when the largest data rate is achieved, the altitude is 175 m and the horizontal distance is 80 m.

Based on these simulation results, we can conclude that there exists an optimal UAV's height, and an optimal horizontal distance between the UAV relay and the BS to achieve the highest data rate.

2. **UAV BS**: As we have mentioned before, fixed-wing and rotary-wing UAVs can both serve as the BS.

**Fixed-Wing UAV BS** As shown in Fig. 1.9, one fixed-wing UAV works as a BS to serve a UE. We assume that the trajectory of the fixed-wing UAV is a circle parallel to the ground with radius $r$ and height $h_{fix}$. Besides, we denote the horizontal distance between the UE and the circle center by $d_{fix}$. It is assumed that the UAV flies with a constant speed $v$. In addition, we assume that the considered time interval $T$ is much longer than the flight time in a cycle, i.e., $T \gg \frac{2\pi r}{v}$. Therefore, the

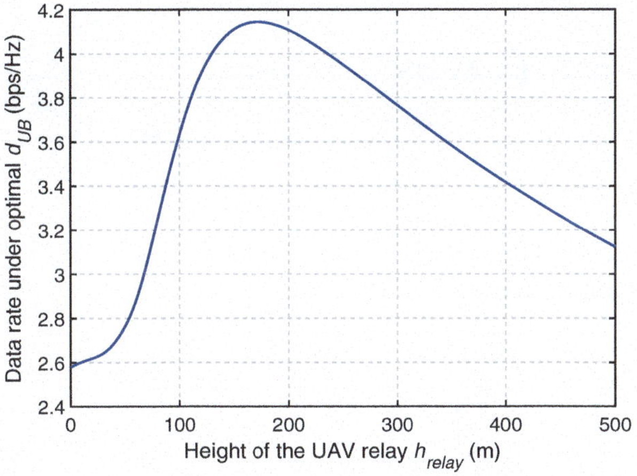


**Fig. 1.7** Data rate under optimal $d_{UB}$ vs. height of the UAV relay $h_{relay}$

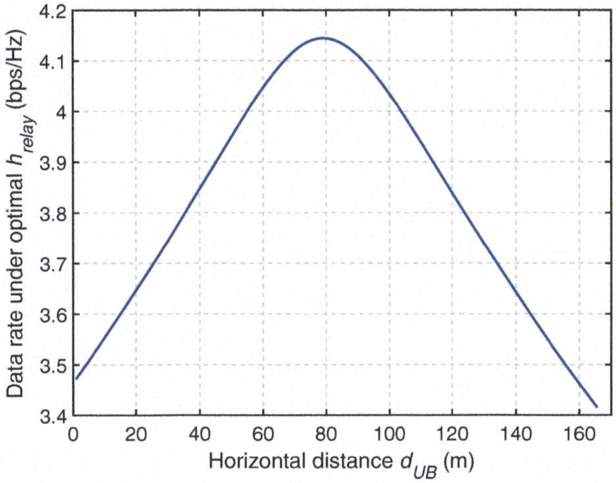


**Fig. 1.8** Data rate under optimal $h_{relay}$ vs. horizontal distance $d_{UB}$

average data rate of the UAV over the time interval $T$ will not vary with $T$. The simulation parameters are given in Table 1.4.

Figure 1.10 depicts the downlink data rate versus the height of the UAV. We can observe that there exists one optimal height of the UAV. Moreover, when the terrestrial UE is closer to the projection of the center of the circular trajectory, the average data rate is higher.

**Rotary-Wind UAV BS** As shown in Fig. 1.11, one rotary-wing UAV works as a BS to serve the UE. The UAV is assumed to remain static during the time interval $T$.

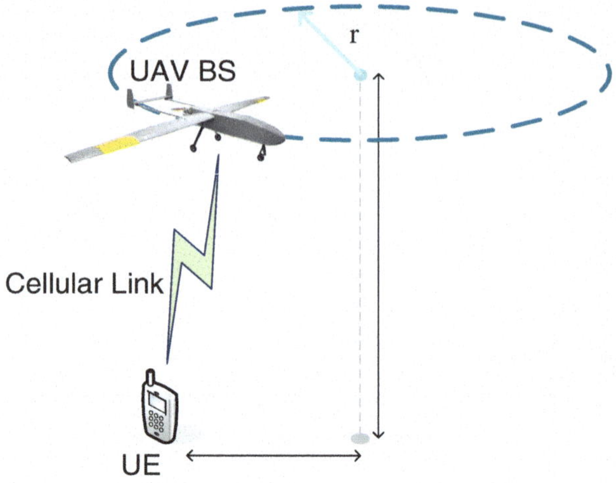


**Fig. 1.9** System model of the UAV BS network with a fixed-wing UAV

**Table 1.4** Simulation parameters for UAV BS

| Parameters | Values |
|---|---|
| Transmit power of the UAV BS | 43 dBm |
| Noise variance | −96 dBm |
| Carrier frequency | 3 GHz |
| The height of the UAV BS $h_{fix}$ | 30 m |
| The height of the UE | 1 m |
| Speed of the fixed-wing UAV $v$ | 18 m/s |
| Trajectory circle radius $r$ | 20 m |
| Antenna pattern of UE | Omnidirection |
| Antenna pattern of UAV BS | Omnidirection |
| Antenna number of UE and UAV BS | 1 Tx, 1 Rx |

For simplicity of discussion, the horizontal distance between the UAV and the UE is denoted by $d_{rot}$, and the height of the UAV is denoted by $h_{rot}$.

Using the same parameters given in Table 1.4, we plot the downlink data rate as the function of the height of the UAV in Fig. 1.12. In the figure, we can find the data rate increases and then decreases as the height of the UAV. We can also observe that the optimal data rate of the rotary-wing UAV BS is larger than that of the fixed-wing UAV BS. This is because the rotary-UAV remains static at the location with the best channel condition while the fixed-wing UAV BS has to move along the trajectory to keep flying.

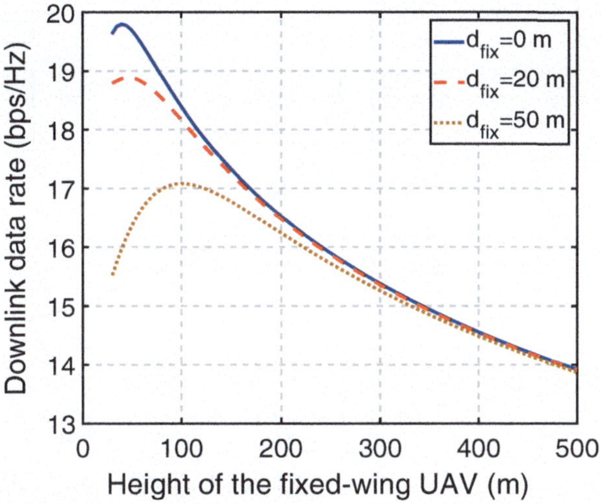


**Fig. 1.10** Downlink data rate vs. height of the UAV

**Fig. 1.11** System model of the UAV BS network with a rotary-wing UAV

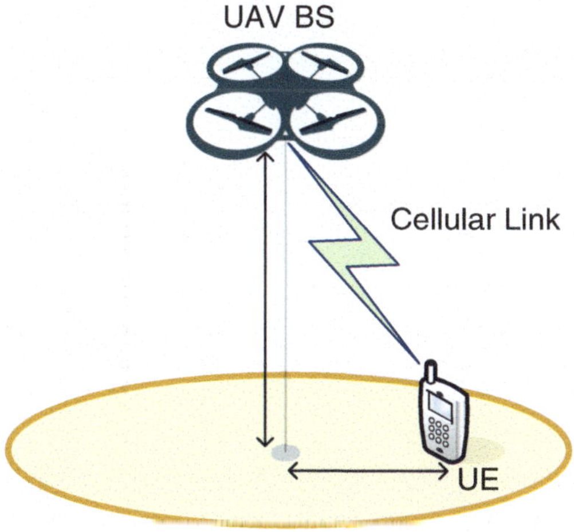


## *1.3.3 Aerial IoT Networks*

In this part, we will analyze the performance of the aerial IoT networks where UAVs serve as aerial users. Here, we assume that the BS is equipped with a linear antenna array, and the antenna array is vertical to the ground, as shown in Fig. 1.13. To better describe the antenna gain, we establish a Cartesian coordinate system, where the antenna is set as the origin $O$, and the z-axis and the x-axis are vertical to the ground plane and the antenna panel, respectively. Given direction $\mathbf{d}$, define $\theta$ as the angle between z-axis and $\mathbf{d}$, and define $\phi$ as the angle between the x-axis and

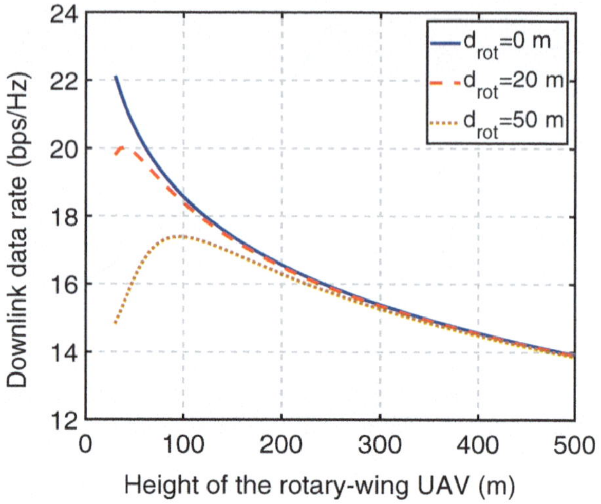


**Fig. 1.12** Downlink data rate vs. height of the UAV

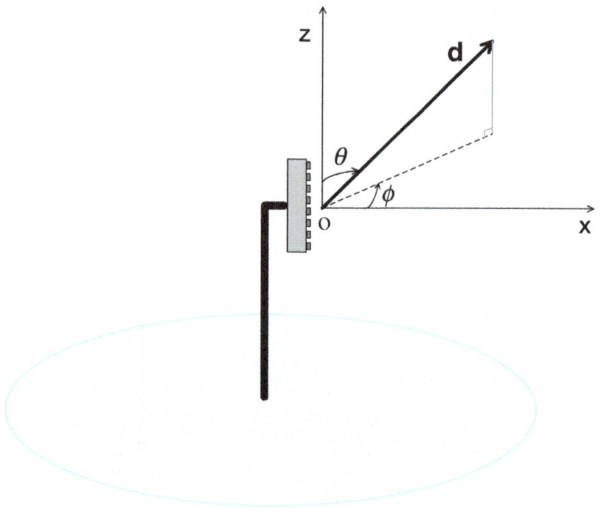


**Fig. 1.13** Schematic diagram of the antenna

the projection of **d** on the x$O$y plane. It can be found that the direction **d** can be described with $\theta$ and $\phi$.

It is assumed that the antenna array is composed of 8 cross-polarized antenna elements, where the separation between every two adjacent antenna elements is half wavelength. Besides, we use the antenna pattern proposed in [17], and assume that the antenna has a tilt angle of 10°.

In Fig. 1.14, we give the antenna gain versus the angle $\theta$, where the angle $\phi$ is set as 0°. From Fig. 1.14, we can observe that

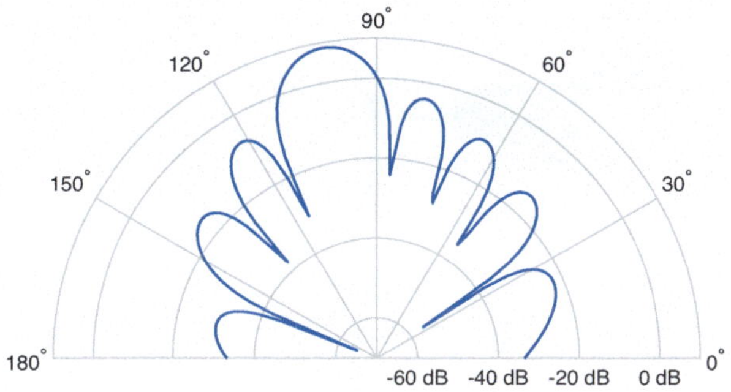


**Fig. 1.14** Antenna gain vs. $\theta$, with $\phi = 0°$

**Table 1.5** Simulation parameters for aerial users

| Parameters | Values |
|---|---|
| Cell radius | 250 m |
| Carrier frequency | 3 GHz |
| Transmit power of the BS | 43 dBm |
| Transmit power of UAV UEs | 23 dBm |
| Noise variance | −96 dBm |
| Antenna pattern of UAV UE | Omnidirection |
| Antenna number of UAV UE and BS | 1 Tx, 1 Rx |

1. The main lobe is directing downwards with a tilt angle of 10°.
2. There are several side lobes with smaller gains.

Simulation parameters are given in Table 1.5.

1. *Single Cell*: As shown in Fig. 1.15, we consider the downlink of a single cell network, where one UAV UE is served by a BS equipped with directional antennas.

It is assumed that the UAV is placed directly opposite to one antenna of the BS. For simplicity of discussion, the horizontal distance between the UAV UE and the BS is denoted by $d_{UB}$ and the height of the UAV UE is denoted by $h_{UAV}$.

In Fig. 1.16, we plot the downlink data rate versus the height of the UAV UE, from which we can observe that

1. The change of the downlink data rate is significant, with the maximum rate around 10 bps/Hz and the minimum rate around 0 bps/Hz, which is caused by the change of the BS antenna gain.
2. Define the minimum height as the UAV height where the local minimum data rate is achieved. According to Fig. 1.16, multiple minimum heights can be found. Besides, when the horizontal distance between the UAV UE and the BS increases, the interval between two adjacent minimum heights becomes larger.

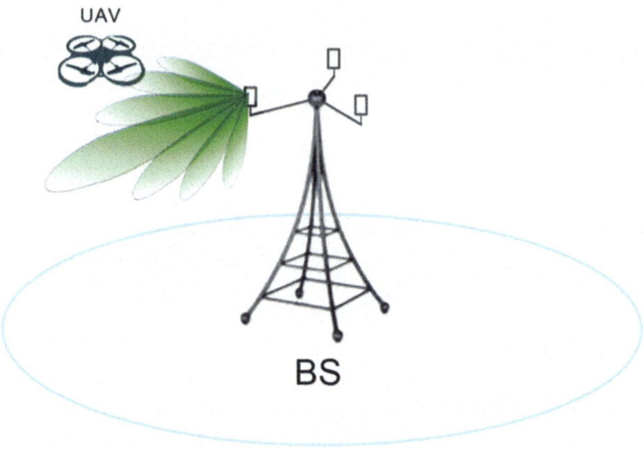


**Fig. 1.15** System model of the single cell network with UAV UE

**Fig. 1.16** Downlink data rate vs. the height of the UAV UE

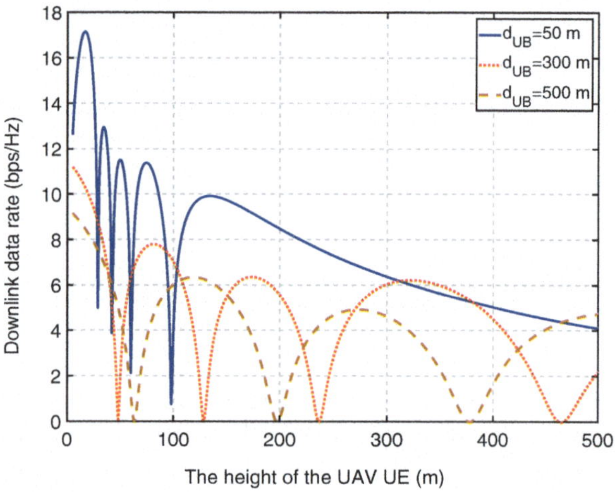


In Fig. 1.17, we depict the downlink data rate versus the horizontal distance between the UAV UE and the BS $d_{UB}$. From Fig. 1.17, we can find that when the downlink horizontal distance increases, the data rate does not decrease strictly, which is caused by the fluctuation of the BS antenna gain.

Based on the simulation results, we can conclude that since the downlink data rate does not strictly decrease with the horizontal distance between the UAV UE and the BS increases, the association of the UAV UE with the BS does not only depend on the horizontal distance.

2. *Three Cells*: To evaluate the interference from aerial UE to other cells, we consider a 3-cell uplink network which provides supports to both aerial UEs and traditional terrestrial UE, as shown in Fig. 1.18. The three cells are denoted by Cell 1, 2, and 3, respectively. It is assumed that Cell 1 contains one aerial UE,

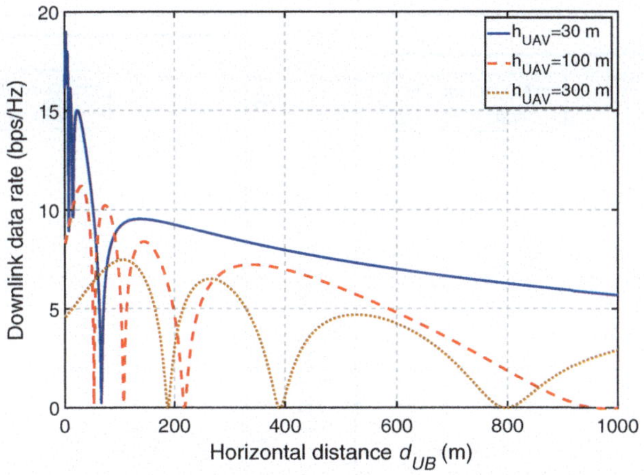


**Fig. 1.17** Downlink data rate vs. horizontal distance $d_{UB}$

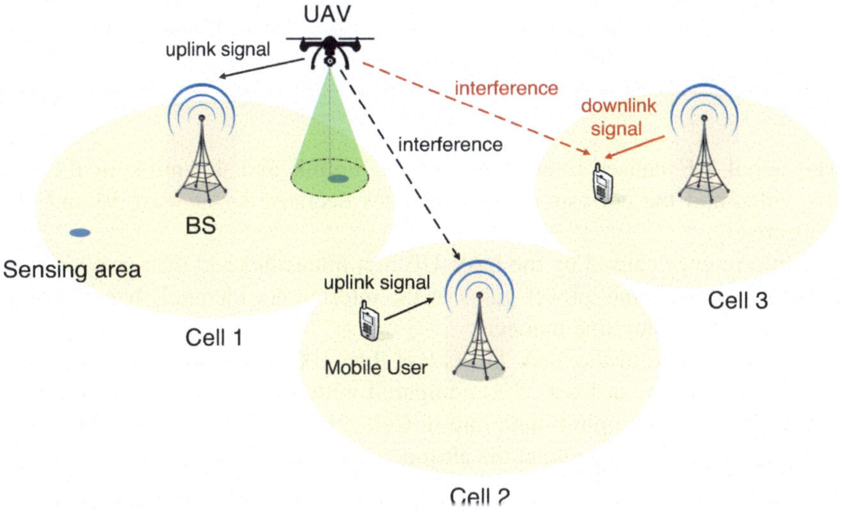


**Fig. 1.18** System model of 3-cell network for uplink aerial UEs

and both Cell 2 and Cell 3 contain one terrestrial UE. To evaluate the interference from UAV UE to the uplinks and downlinks in other cells, we consider the uplink in Cell 1 and Cell 2, and the downlink in Cell 3.

Figure 1.19 depicts the data rate in different cells versus height of UAV UE. To verify the interference that the UAV UE causes to the BS and the terrestrial UE, the scenario without UAV UE is also considered. According to Fig. 1.19, we can observe that

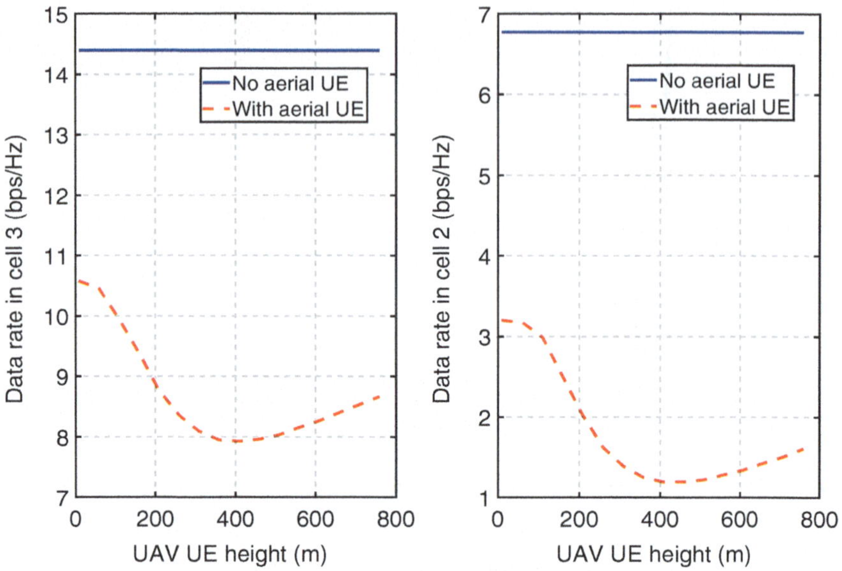


**Fig. 1.19** Data rate vs. height of the uplink aerial UE

1. The aerial UE causes interference to the uplink and downlink in the other two cells, and the corresponding data rates decrease by at least 50 and 30%, respectively.
2. The interference caused by the UAV UE first intensifies and then weakens. This is because the channel power gains of the interference channel change with the UAV altitude in the same manner.
3. When the altitude of the UAV UE is less than 100 m, the downlink data rate in Cell 3 increases by at least 25%, compared with the data rate at the altitude of 400 m. Besides, the uplink data rate in Cell 2 increases by at least 150% when compared with the data rate at the altitude of 400 m.

To evaluate the interference to UAV UE from other cells, we also consider a 3-cell downlink network, as shown in Fig. 1.20. The three cells are denoted by Cell 1, 2, and 3, respectively. It is assumed that Cell 1 contains one UAV UE, and both Cell 2 and Cell 3 contain one terrestrial UE. To evaluate the interference to UAV UE from the links in other cells, downlinks are assumed for both Cell 1 and Cell 2, and uplink is assumed for Cell 3.

Figure 1.21 depicts the data rate of UAV UE versus the height of UAV UE. To verify the interference that other cells causes to the UAV UE, we also consider a single cell case which contains one UAV UE. From Fig. 1.21, we can observe that

1. Links in Cell 2 and Cell 3 cause interference to the UAV UE, and the downlink data rate of the UAV UE decreases by at least 35%.

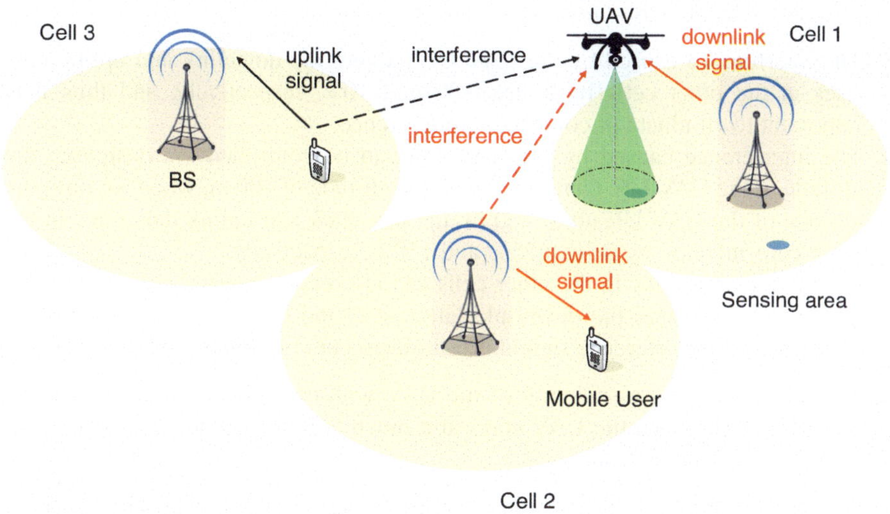


**Fig. 1.20** System model of 3-cell network with downlink aerial UE

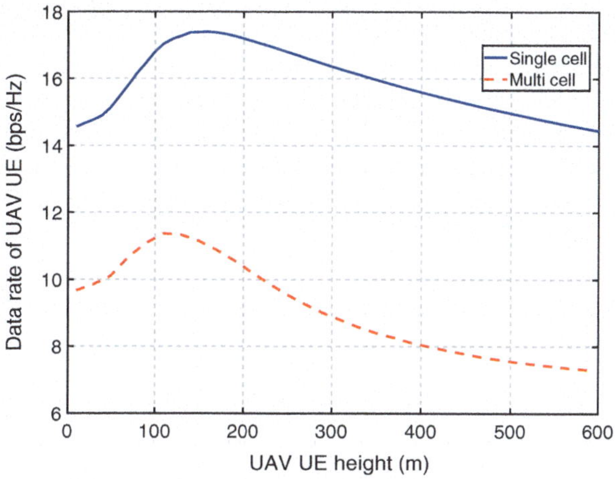


**Fig. 1.21** Data rate of UAV UE vs. height of the downlink aerial UE

2. When the UAV altitude grows, the data rate of the UAV in the multi-cell case increases first, and then decreases. This is because the channel gain from the serving BS to the UAV UE first increases and then decreases. Moreover, we can find that the height with the optimal data rate is achieved in the 3-cell case is different from that in the single cell case, due to the interference to the UAV UE.

Based on the simulation results, we can conclude that

i. The interference caused by the UAV UE reduces the downlink and uplink data rates in the other cells by at least 30 and 50%, respectively, and thus it is important to eliminate or control the interference.
ii. The interference caused by the UAV UE can be controlled via designing the altitude of the UAV UE. Compared with the altitude of 400 m, when we limit the altitude of the UAV UE up to 100 m, the downlink and uplink data rates in the other cells increase by at least 25% and 150%, respectively.
iii. Communication links in the other cells can also cause interference to the UAV UE, which decreases the downlink data rate of the UAV UE by at least 35%. Therefore, the interference should be eliminated or controlled.

3. *Multiple Cells*: The association of the UAV with the BSs is different from the terrestrial UEs, since the UAV stays at a much higher altitude. In this part, we study the association issue.

As shown in Fig. 1.22, we consider a multi-cell network which provides support to UAV UEs. For simplicity of discussion, we define set $\mathcal{L}_1$ as the cell in the center, and define set $\mathcal{L}_i$ $(i > 1)$ as the set of cells adjacent to the cells in set $\mathcal{L}_{i-1}$. It is assumed that the outermost cells belong to set $\mathcal{L}_5$. Besides, all the BSs are equipped with the directional antennas.

We focus on only one UAV UE, whose horizontal location obeys uniform distribution in the $\mathcal{L}_1$ cell. Besides, the UAV UE chooses to associate with the BS that provides the maximum received power. For simplicity of discussion, $d_{asso}$ is used to represent the horizontal distance between the UAV UE and the BS that it associates with, and the height of the UAV UE is denoted by $h_{UAV}$. The simulation parameters are given in Table 1.5.

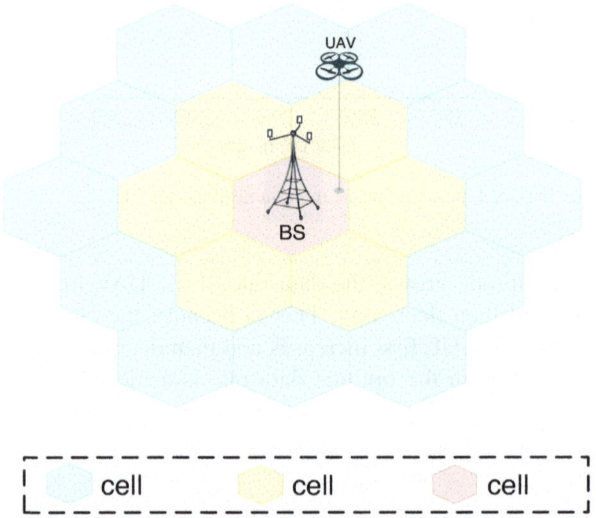


**Fig. 1.22** System model of multi-cell network

Figure 1.23 plots the cumulative distribution function (CDF) of the horizontal distance from the UAV to the BS that it associates with. We can observe that the CDF curve can be divided into three parts, according to the monotonicity. Specifically, the CDF curve first increases with the horizontal distance. Then the curve remains unchanged. After that, the curve increases again. Denote the horizontal distances corresponding to the two end points of the second part of the CDF curve by $d_l$ and $d_r$, respectively. The CDF curve remains unchanged when the horizontal distance between a BS and the UAV UE is in the interval $[d_l, d_r]$, as the antenna gain of the BS is small. In addition, we can find that $d_l$ and $d_r$ are influenced by the height of the UAV.

Based on the simulation results, we can conclude that the UAV UE may not associate with the closest BS.

### *1.3.4 Propulsion and Mobility Model*

In this book, we assume that the trajectory of a UAV is a line segment in two successive time slots, as shown in Fig. 1.24. Define $\boldsymbol{x}(t)$ as the 3D location of the UAV at time slot $t$. Therefore, the velocity $\boldsymbol{v}(t)$ can be defined as

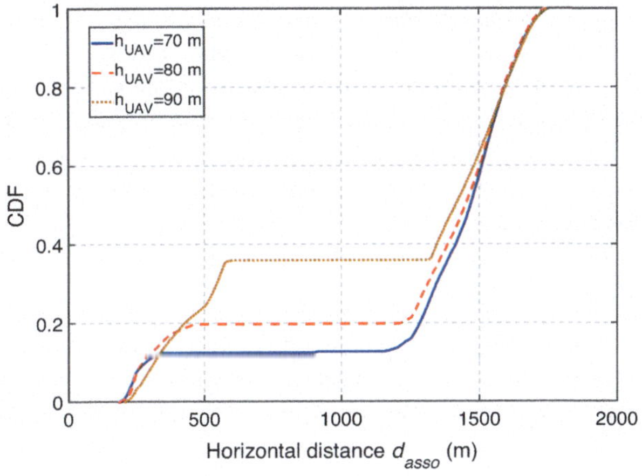


**Fig. 1.23** CDF of horizontal distance $d_{asso}$

**Fig. 1.24** Trajectory of a UAV

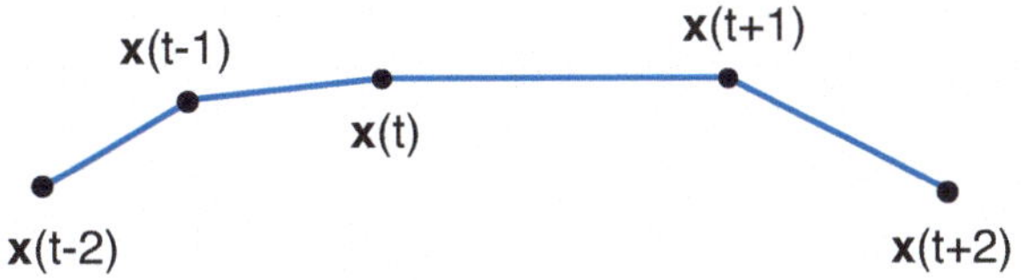

**Fig. 1.25** Forces on a UAV

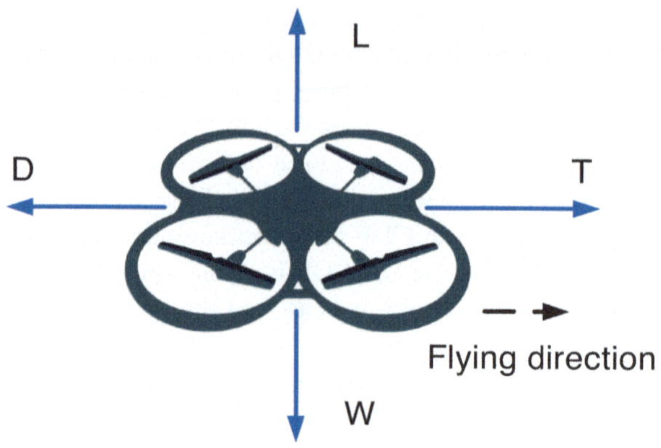


$$\boldsymbol{v}(t) = \boldsymbol{x}(t) - \boldsymbol{x}(t-1), \tag{1.7}$$

and the accelerate velocity $\boldsymbol{a}(t)$ can be defined as

$$\boldsymbol{a}(t) = \boldsymbol{v}(t) - \boldsymbol{v}(t-1). \tag{1.8}$$

Due to the space and mechanical limitations, the UAV has a maximum flight altitude $h_{max}$, a maximum velocity $v_{max}$, and a maximum accelerated velocity $a_{max}$, that is, $H_{UT} \leq h_{max}$, $\|\boldsymbol{v}(t)\| \leq v_{max}$, and $\|\boldsymbol{a}(t)\| \leq a_{max}$.

Under the mobility model, the total energy consumption of the UAV includes two components. The first one is the communication-related energy, which is due to the radiation and signal processing. The other component is the propulsion energy, which is required to support its mobility. In practice, the communication-related energy is usually less than the propulsion one. In the following, we will introduce the propulsion energy model.

As shown in Fig. 1.25, an aircraft aloft is in general subject to four forces[1]: weight, drag, lift, and thrust [18]:

- Weight (W): the force of gravity which is related with the aircraft's mass $M$.
- Drag (D): the aerodynamic force component parallel to the flying direction.
- Lift (L): the aerodynamic force component normal to the drag and pointing upward.
- Thrust (F): the force produced by the aircraft engine, which overcomes the drag to move the aircraft forward.

Assume that the UAV moving at a speed $v$, the drag can be expressed by Leishman [19]

[1]Since both fixed-wing and rotary-wing UAVs will receive these four forces, this model can be used by any one type of UAVs.

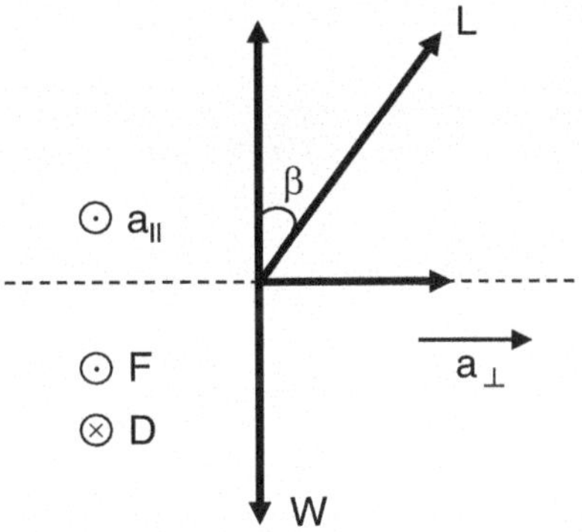


**Fig. 1.26** Illustrations on the forces to accelerate the UAV. The UAV flies normal to the page

$$D = \kappa_1 v^2 + \frac{\kappa_2 \mu}{v^2}, \tag{1.9}$$

where $\kappa_1$ and $\kappa_2$ are constants related to the weight $W$ and the air environment, and $\mu = \frac{L}{W}$ is the load factor.

For a UAV to change its flying direction, the lift $L$ will generate a horizontal force to support the centrifugal acceleration, i.e., the acceleration component normal to the velocity. As illustrated in Fig. 1.26, let $\beta$ denote the angle between the vertical plane and the direction of the lift. We then have the following equations:

$$L \cos \beta = W, \tag{1.10}$$

$$L \sin \beta = Ma_{\perp}, \tag{1.11}$$

$$F - D = Ma_{\parallel}, \tag{1.12}$$

where $a_{\perp}$ and $a_{\parallel}$ denote the acceleration components that are perpendicular and parallel to the velocity, respectively, with $a > 0$ for accelerating and $a < 0$ for decelerating. According to (1.10) and (1.11), we can have

$$\tan \beta = \frac{a_{\perp}}{g}. \tag{1.13}$$

Therefore, we can derive that $\mu = 1 + \tan^2 \beta = 1 + \frac{a_{\perp}^2}{g^2}$.

Moreover, according to (1.9) and (1.13), we can have

$$F = \kappa_1 v^2 + \frac{\kappa_2}{v^2}\left(1 + \frac{a_{\perp}^2}{g^2}\right) + Ma_{\parallel}. \tag{1.14}$$

Therefore, the required instantaneous power can be given by

$$P_R = |F|v = |\kappa_1 v^3 + \frac{\kappa_2}{v}\left(1 + \frac{a_\perp^2}{g^2}\right) + Ma_\| v|. \tag{1.15}$$

Note that the acceleration $\boldsymbol{a}$ can be decomposed along the parallel and normal directions of $\boldsymbol{v}$, that is,

$$a_\| = \frac{\boldsymbol{a}^T \boldsymbol{v}}{\|\boldsymbol{v}\|_2}, \quad a_\perp = \sqrt{\|\boldsymbol{a}\|_2^2 - \frac{(\boldsymbol{a}^T \boldsymbol{v})^2}{\|\boldsymbol{v}\|_2^2}}. \tag{1.16}$$

Thus, the total propulsion energy can be calculated by

$$\begin{aligned} E_{prop} = & \int_{t=0}^{T} \left( \kappa_1 \|\boldsymbol{v}(t)\|_2^3 + \frac{\kappa_2}{\|\boldsymbol{v}(t)\|_2} \left( 1 + \frac{\|\boldsymbol{a}(t)\|_2^2 - \frac{(\boldsymbol{a}^T(t)\boldsymbol{v}(t))^2}{\|\boldsymbol{v}(t)\|_2^2}}{g^2} \right) \right) \mathrm{d}t \\ & + \int_{t=0}^{T} M\boldsymbol{a}^T(t)\boldsymbol{v}(t)\mathrm{d}t. \end{aligned} \tag{1.17}$$

Further, we have

$$\begin{aligned} \int_{t=0}^{T} M\boldsymbol{a}^T(t)\boldsymbol{v}(t)\mathrm{d}t &= \int_{t=0}^{T} M\frac{\boldsymbol{v}^T(t)}{dt}\boldsymbol{v}(t)\mathrm{d}t \\ &= \tfrac{1}{2}M\left(\|\boldsymbol{v}(T)\|_2^2 - \|\boldsymbol{v}(0)\|_2^2\right). \end{aligned} \tag{1.18}$$

Therefore, the total propulsion energy can be bounded by

$$\begin{aligned} E_{prop} \le & \int_{t=0}^{T} \left( \kappa_1 \|\boldsymbol{v}(t)\|_2^3 + \frac{\kappa_2}{\|\boldsymbol{v}(t)\|_2} \left( 1 + \frac{\|\boldsymbol{a}(t)\|_2^2}{g^2} \right) \right) \mathrm{d}t \\ & + \tfrac{1}{2}M\left(\|\boldsymbol{v}(T)\|_2^2 - \|\boldsymbol{v}(0)\|_2^2\right). \end{aligned} \tag{1.19}$$

Since $\frac{1}{2}M\left(\|\boldsymbol{v}(T)\|_2^2 - \|\boldsymbol{v}(0)\|_2^2\right)$ is constant, in the optimization problems in the following chapters, we omit this term for brevity.

It is worthwhile to point out that it is an approximation when the speed of the UAV $v$ is larger than a threshold $v_{th}$, where

$$v_{th} = \sqrt{\frac{M}{2A\rho}}. \tag{1.20}$$

Here, $A$ is the rotor disc area and $\rho$ is the air density. When the speed is lower than $v_{th}$, the energy consumption can be regarded as a constant.

## References


1. L. Gupta, R. Jain, G. Vaszkun, Survey of important issues in UAV communication networks. IEEE Commun. Surv. Tutorials **18**(2), 1123–1152 (2016)
2. I.K. Nikolos, K.P. Valavanis, N.C. Tsourveloudis, A.N. Kostaras, Evolutionary algorithm based offline/online path planner for UAV navigation. IEEE Trans. Syst. Man Cybern. B Cybern. **33**(6), 898–912 (2003)
3. X. Wang, V. Yadav, S.N. Balakrishnan, Cooperative UAV formation flying with obstacle/collision avoidance. IEEE Trans. Control Syst. Technol. **15**(4), 672–679 (2007)
4. ITU-R, IMT vision – framework and overall objectives of the future development of IMT for 2020 and beyond, M.2083-0 (2015)
5. M. Hung, Leading the IoT, Gartner insights on how to lead in a connected world. https://www.gartner.com/imagesrv/books/iot/iotEbook_digital.pdf
6. H. Zhang, L. Song, Z. Han, H.V. Poor, Cooperation techniques for a cellular internet of unmanned aerial vehicles. IEEE Wireless Commun. **26**(5), 167–173 (2019)
7. A. Fotouhi, H. Qiang, M. Ding, M. Hassan, L.G. Giordano, A.G.-Rodriguez, J. Yuan, Survey on UAV cellular communications: practical aspects, standardization advancements, regulation, and security challenges. IEEE Commun. Surv. Tutorial, Early access, Arxiv: https://arxiv.org/abs/1809.01752
8. Nokia, F-cell technology from Nokia Bell Labs revolutionizes small cell deployment by cutting wires, costs and time (2016). https://www.nokia.com/news/releases/2016/10/03/f-cell-technology-from-nokia-bell-labs-revolutionizes-small-cell-deployment-by-cuttingwires-costs-and-time/
9. Facebook, Building communications networks in the stratosphere. https://code.facebook.com/posts/993520160679028/building-communications-networks-in-the-stratosphere/
10. ITU-R, Characteristics of unmanned aircraft systems and spectrum requirements to support their safe operation in non-segregated airspace, M.2171 (2009)
11. S. Zhang, J. Yang, H. Zhang, L. Song, Dual trajectory optimization for a cooperative internet of UAVs. IEEE Commun. Lett. **23**(6), 1093–1096 (2019)
12. N. Ahmed, S.S. Kanhere, S. Jha, On the importance of link characterization for aerial wireless sensor networks. IEEE Commun. Mag. **54**(5), 52–57 (2016)
13. A. Al-Hourani, S. Kandeepan, A. Jamalipour, Modeling air-to-ground path loss for low altitude platforms in urban environments, in *Proceedings of IEEE GLOBECOM,* Austin (2014)
14. 3GPP, Enhanced LTE supported for aerial vehicles, TR 36.777, V 15.0.0 (2018)
15. D.W. Matolak, R. Sun, Air-ground channel characterization for unmanned aircraft system-part i: methods, measurements, and models for over-water settings. IEEE Trans. Veh. Technol. **66**(1), 26–44 (2017)
16. 3GPP, Study on channel model for frequencies from 0.5 to 100 GHz, TR 38.901, V 15.0.0 (2018)
17. 3GPP, Study on 3D channel model for LTE, TR 36.873 V 12.7.0 (2017)
18. Y. Zeng, R. Zhang, Energy efficient UAV communication with trajectory optimization. IEEE Trans. Wireless Commun. **16**(6), 3747–3760 (2017)
19. G.J. Leishman, *Principles of Helicopter Aerodynamics* (Cambridge University Press, Cambridge, 2006), pp. 159–193
20. Z. Hu, Z. Zheng, L. Song, T. Wang, X. Li, UAV offloading: spectrum trading contract design for UAV-assisted cellular networks. IEEE Trans. Wireless Commun. **17**(9), 6093–6107 (2018)
21. S. Zhang, H. Zhang, Q. He, K. Bian, L. Song, Joint power and trajectory optimization for UAV relay networks. IEEE Commun. Lett. **22**(1), 161–164 (2018)

# Chapter 2
# Basic Theoretical Background

In this chapter, we will review some theories which will be used in the following chapters, including optimization theory, game theory, and machine learning.

This chapter is organized as follows: We give some brief introductions of optimization theory in Sect. 2.1. In Sect. 2.2, we give the overview of game theory. In Sect. 2.3, the related machine learning technologies are presented.

## 2.1 Brief Introductions to Optimization Theory

In this section, we will discuss how to solve the following optimization problem to find the minimum of $f$ over $\mathscr{X}$, i.e.,

$$\begin{aligned} &\min f(\boldsymbol{x}) \\ &\text{s.t. } \boldsymbol{x} \in \mathscr{X}. \end{aligned} \tag{2.1}$$

According to the types of the variables, i.e., continuous and integer, the optimization problems can be categorized into two types[1]: continuous and integer optimization problems. In the following, we will elaborate on how to solve these two types of optimization problems, respectively.

[1]There also exists a mixed type of optimization problems, which consist of both continuous and integer variables. However, some existing approaches can decompose this type of problems into two subproblems, which contain only continuous or integer variables. Therefore, in this section, we only introduce how to solve these two basic optimization problems.



H. Zhang et al., *Unmanned Aerial Vehicle Applications over Cellular Networks for 5G and Beyond*, Wireless Networks, https://doi.org/10.1007/978-3-030-33039-2_2

## 2.1.1 Continuous Optimization

In continuous optimization problems, convexity is an important concept because the local optimal solution of a convex function is also the global optimal solution. Therefore, this characteristic of convex functions allows for some algorithms to optimize them with a quick convergence speed [1].

Formally, a set $\mathscr{X} \subset \mathbb{R}^d$ is said to be *convex* if it contains all of its segments, that is

$$\forall(x, y, \gamma) \in \mathscr{X} \times \mathscr{X} \times [0, 1], (1 - \gamma)x + \gamma y \in \mathscr{X}. \tag{2.2}$$

A function $f : \mathscr{X} \to \mathbb{R}$ is said to be convex if it always lies below its chords, that is

$$\forall(x, y, \gamma) \in \mathscr{X} \times \mathscr{X} \times [0, 1], f((1-\gamma)x+\gamma y) \leq (1-\gamma)f(x)+\gamma f(y). \tag{2.3}$$

A problem is called convex if both objective and constraints of the problem are convex. In this part, we will discuss how to solve convex and non-convex optimization problems, respectively.

### 2.1.1.1 Convex Optimization Problem

Generally, a convex optimization problem can be given by

$$\begin{aligned} &\min f_0(\boldsymbol{x}) \\ \text{s.t. } &f_i(\boldsymbol{x}) \leq 0, i = 1, \ldots, m \\ &h_j(\boldsymbol{x}) = 0, j = 1, \ldots, p, \end{aligned} \tag{2.4}$$

where $f_0, \ldots, f_m, h_1, \ldots, h_p$ are convex and twice continuously differentiable. We assume that the problem has at least one optimal solution $\boldsymbol{x}^*$. Define $\mathscr{X}$ is the domain of the problem, we also assume that the problem is strictly feasible, i.e., $\mathscr{X}$ is not empty.

**Duality** Before the algorithm to solve this problem, we first introduce the concept of *duality*. The basic idea for Lagrange duality is to take the constraints in (2.4) into account by augmenting the objective function with a weighted sum of the constraint functions. Lagrange function $L$ associated with the problem can be written by

$$L(\boldsymbol{x}, \boldsymbol{\lambda}, \boldsymbol{\mu}) = f_0(\boldsymbol{x}) + \sum_{i=1}^{m} \lambda_i f_i(\boldsymbol{x}) + \sum_{j=1}^{p} \mu_i h_i(\boldsymbol{x}). \tag{2.5}$$

We refer to $\lambda_i$ as the Lagrange multiplier associated with the $i$-th inequality constraint $f_i(\boldsymbol{x}) \leq 0$ and $\mu_j$ as the Lagrange multiplier associated with the $j$-th equality constraint $h_j(\boldsymbol{x}) = 0$. We also define the *dual function* $g$ as the minimum value of the Lagrangian over $\boldsymbol{x}$

$$g(\boldsymbol{\lambda}, \boldsymbol{\mu}) = \min_{\boldsymbol{x} \in \mathscr{X}} L(\boldsymbol{x}, \boldsymbol{\lambda}, \boldsymbol{\mu}). \tag{2.6}$$

According to the results in [2], for a generalized problem given in the form of (2.4), the dual function yields the lower bound on the optimal value, but they are equal when the problem is convex.

**Primal-Dual Methods** With the help of the dual function, we can obtain the optimal solution by solving the dual problem, which is given by

$$\max_{\boldsymbol{\lambda}, \boldsymbol{\mu} \succeq 0} g(\boldsymbol{\lambda}, \boldsymbol{\mu}). \tag{2.7}$$

We can solve this problem iteratively by decomposing it into two subproblems. The master subproblem is the optimization of $\boldsymbol{\lambda}$ and $\boldsymbol{\mu}$ and the slave subproblem is to optimize $\boldsymbol{x}$ given $\boldsymbol{\lambda}$ and $\boldsymbol{\mu}$. In the following, we will discuss them in detail.

1. *Slave Subproblem*: The slave subproblem is a non-constrained convex optimization problem, and thus we can obtain the solution by equaling the first derivative of the Lagrange function $L(\boldsymbol{x}, \boldsymbol{\lambda}, \boldsymbol{\mu})$ over $\boldsymbol{x}$ to 0, i.e. $\nabla_{\boldsymbol{x}} L(\boldsymbol{x}, \boldsymbol{\lambda}, \boldsymbol{\mu}) = 0$.

   However, for some functions, we may not obtain the solutions analytically, and thus we can use the gradient method instead. Specially, we can start from an initial solution $\boldsymbol{x}^0 \in \mathscr{X}$. In the $l$-th iteration, we can calculate the gradient $\nabla_{\boldsymbol{x}} L(\boldsymbol{x}, \boldsymbol{\lambda}, \boldsymbol{\mu})_{\boldsymbol{x}=\boldsymbol{x}^{l-1}}$, and then update the solution with step size $\eta_i^l$ for variable $x_i$ according to the following rule:

$$x_i^l = x_i^{l-1} - \eta_i^l (\nabla_{\boldsymbol{x}} L(\boldsymbol{x}, \boldsymbol{\lambda}, \boldsymbol{\mu})_{\boldsymbol{x}=\boldsymbol{x}^{l-1}})_i, \forall i. \tag{2.8}$$

   It is worthwhile to point out that the step size $\eta_l$ should decrease as $l$ grows. When the solution is far from the optimal one, the step size can be large to accelerate the convergence speed. However, when the solution is close to the optimal one, the step size should be smaller to get a higher accuracy.
2. *Master Subproblem*: In the master subproblem, we can also update the Lagrange multipliers $\boldsymbol{\lambda}$ and $\boldsymbol{\mu}$ by the gradient method, as given by

$$\begin{aligned} \lambda_i^l &= \lambda_i^{l-1} - \delta_i^l \left[f_i(\boldsymbol{x})\right]^+, \forall i \\ \mu_j^l &= \mu_j^{l-1} - \theta_j^l \left[h_j(\boldsymbol{x})\right]^+, \forall j. \end{aligned} \tag{2.9}$$

   Here, $\delta_i^l$ is the step size for $\lambda_i$ in the $l$-th iteration, $\theta_j^l$ is the step size for $\mu_j$ in the $l$-th iteration, and $[x] = \max\{0, x\}$.

**Optimal Conditions** Let $\boldsymbol{x}^*$ and $(\boldsymbol{\lambda}^*, \boldsymbol{\mu}^*)$ be any primal and dual optimal points. Due to the convexity of the problem, the primal and dual optimal points are the same. Since $\boldsymbol{x}^*$ minimizes $L(\boldsymbol{x}^*, \boldsymbol{\lambda}^*, \boldsymbol{\mu}^*)$ over $\boldsymbol{x}$, its gradient must vanish at $\boldsymbol{x}^*$, i.e.,

$$\nabla f_0(\boldsymbol{x}^*) + \sum_{i=1}^{m} \lambda_i^* \nabla f_i(\boldsymbol{x}^*) + \sum_{j=1}^{p} \mu_i^* \nabla h_i(\boldsymbol{x}^*) = 0. \tag{2.10}$$

Thus, we have the following equations:

$$\begin{aligned} & f_i(\boldsymbol{x}^*) \leq 0, \\ & h_j(\boldsymbol{x}^*) = 0, \\ & \lambda_i^* \geq 0, \\ & \lambda_i^* f_i(\boldsymbol{x}^*) = 0, \\ & \nabla f_0(\boldsymbol{x}^*) + \sum_{i=1}^{m} \lambda_i^* \nabla f_i(\boldsymbol{x}^*) + \sum_{j=1}^{p} \mu_i^* \nabla h_i(\boldsymbol{x}^*) = 0. \end{aligned} \tag{2.11}$$

These equations are called Karush–Kuhn–Tucker (KKT) conditions, which provide the necessary conditions for the convex optimizations. KKT conditions are important in optimization theory because KKT conditions can be solved analytically in a few special cases, and thus the solutions of the problem can be derived directly.

#### 2.1.1.2 Non-convex Optimization Problem

In a non-convex problem, at least one function in the objective or constraints is non-convex. There does not exit a general-purpose method to obtain the global solution of a non-convex constrained optimization problem. However, for some special kinds of functions, we can use approximation methods to obtain a local optimal solution.

In the context of wireless communications, a widely used non-convex function is the data rate with interference, which can be given by

$$R(\boldsymbol{x}) = \log_2 \left( 1 + \frac{h_1 x_1}{\sigma^2 + \sum_{m=2}^{d} h_m x_m} \right), \tag{2.12}$$

where $\sigma^2$ is a constant denoting the power of the noise, and $h_1, \ldots, h_m$ are also constants denoting the channel gains. However, (2.12) can be rewritten by

$$R(\boldsymbol{x}) = -\log_2 \left( \sigma^2 + \sum_{m=2}^{d} h_m x_m \right) - \left( -\log_2 \left( \sigma^2 + \sum_{m=1}^{d} h_m x_m \right) \right), \tag{2.13}$$

which is the difference of two convex functions. Therefore, in this part, we will focus on how to solve the difference of convex functions (DC) problems [3].

Typically, a DC program[2] can be written by

$$\begin{aligned} \min f(\boldsymbol{x}) &= g(\boldsymbol{x}) - h(\boldsymbol{x}), \\ \text{s.t. } \boldsymbol{x} &\in \mathscr{X}, \end{aligned} \tag{2.14}$$

where $g(\boldsymbol{x})$ and $h(\boldsymbol{x})$ are convex over $\mathscr{X}$, and $\mathscr{X}$ is a convex set.

In the following, we will introduce a DC algorithm to solve the DC program. The basic idea of the DC algorithm is that a stationary point for a difference function occurs when the gradients of the two terms match, i.e.,

$$\nabla g(\boldsymbol{x}) = \nabla h(\boldsymbol{x}). \tag{2.15}$$

The DC algorithm simply turns this premise into an implicit fixed-point scheme, that is,

$$\nabla g(\boldsymbol{x}^{l+1}) = \nabla h(\boldsymbol{x}^l), \tag{2.16}$$

where $\boldsymbol{x}^{l+1}$ is selected so that the gradients match for these two functions at iteration $l$. Although this scheme might look arbitrary, it in fact exploits the features of the function, ensuring a monotonously decreasing sequence.

For an easier implementation, $\boldsymbol{x}^{l+1}$ can be solved by

$$\boldsymbol{x}^{l+1} = \arg \min_{\boldsymbol{x} \in \mathscr{X}} g(\boldsymbol{x}) - \tilde{h}(\boldsymbol{x}_l), \tag{2.17}$$

where $\tilde{h}(\boldsymbol{x}_l)$ is a linear approximation of $h(\boldsymbol{x})$ around $\boldsymbol{x}^l$, i.e.,

$$\tilde{h}(\boldsymbol{x}_l) = h(\boldsymbol{x}^l) + (\boldsymbol{x} - \boldsymbol{x}^l)^T \nabla h(\boldsymbol{x}^l). \tag{2.18}$$

Using this approximation, the objective function is then turned to a simplified convex function, which can be solved by the aforementioned convex techniques.

### 2.1.2 Integer Optimization

In this part, we discuss some basic solution for the following general integer programming problem:

[2]In this section, we only discuss the case where only the objective is DC. When the constraints are DC, we can use the Lagrange method to transform the problem to the one where only the objective is DC.

$$\begin{aligned} \min\ & f_0(\boldsymbol{x}) \\ \text{s.t. } & f_i(\boldsymbol{x}) \le 0, i = 1, \ldots, m, \\ & h_j(\boldsymbol{x}) = 0, j = 1, \ldots, p, \\ & \boldsymbol{x} \in \mathbb{Z}^d, \end{aligned} \tag{2.19}$$

where $f_0, \ldots, f_m, h_1, \ldots, h_p$ are real-valued functions defined on $\mathbb{R}^d$ and $\mathbb{Z}^d$ is the set of integer points in $\mathbb{R}^d$. We define the domain of problem (2.19) as $\mathscr{X}$.

Since the feasible solutions of the problem are discrete, it is straightforward to enumerate all the feasible solutions and select the optimal one. However, the enumeration approach is infeasible for large-scale integer programming problems, and thus the idea of partial enumeration is still attractive. In this subsection, we introduce the branch-and-bound method, which is one of the most widely used partial enumeration schemes [4].

#### 2.1.2.1 Branch-and-Bound Method

The basic idea behind the branch-and-bound method is an implicit enumeration scheme that systematically discards the points in $\mathscr{X}$ that are not possible to achieve the optimality. To partition the search space, we divide the integer set $\mathscr{X}$ into $k (k \ge 2)$ subsets: $\mathscr{X}_1, \ldots, \mathscr{X}_k$. A subproblem at node $i$ is formed by replacing $\mathscr{X}$ with $\mathscr{X}_i$, and these subproblems form a subproblem list $\mathscr{L}$.

For selected node $i$, a lower bound $B_i$ for the optimal value of the subproblem is estimated. If $B_i$ is greater than or equal to the best feasible solution found, then the subproblem is fathomed from further consideration. Otherwise, this subproblem is kept in the subproblem list. For one unfathomed node $\mathscr{X}_i$, it will be further branched into smaller subsets. The process is repeated until there is no subproblem left in the list. It is helpful to use a tree structure to describe the branch-and-bound method where a node stores the information necessary for solving the corresponding subproblem. The steps of the branch-and-bound method are elaborated on below:

1. *Initialization*: Set the subproblem list $\mathscr{L}$. Set an initial feasible solution as the best one $\boldsymbol{x}^*$ and its corresponding objective value $f_0(\boldsymbol{x}^*)$. If no feasible solution is found, then set $f_0(\boldsymbol{x}^*) = +\infty$.
2. *Node Selection*: If $\mathscr{L}$ is empty, stop and $\boldsymbol{x}^*$ is the optimal solution. Otherwise, select $k$ nodes from $\mathscr{L}$. Denote the set of selected nodes by $\mathscr{L}^s$ and use $\mathscr{L} \backslash \mathscr{L}^s$ to replace $\mathscr{L}$. Set $i = 1$.
3. *Bounding*: Compute a lower bound $B_i$ of the subproblem at node $i$. Set $B_i = +\infty$ if the problem is infeasible. If $B_i > f_0(\boldsymbol{x}^*)$, go to the fathoming step.
4. *Feasible Solution*: Save the best feasible solution found in the bounding step or generate a better feasible solution by a certain heuristic method. Update $\boldsymbol{x}^*$ and $f_0(\boldsymbol{x}^*)$ when the feasible solution is better than the current one. Remove subproblem at node $j$ from $\mathscr{L}^s$ when $B_j > f_0(\boldsymbol{x}^*)$. If $i < k$, set $i = i + 1$ and return to the bounding step. Otherwise, go to the branching step.

5. *Branching*: If $\mathscr{L}$ is empty, go to the node selection step. Otherwise, choose a node $i$ from $\mathscr{L}^s$. Further divide node $i$ into several smaller subsets $\mathscr{L}^b$. Remove node $i$ from $\mathscr{L}^s$ and set $\mathscr{L} = \mathscr{L} \cup \mathscr{L}^s \cup \mathscr{L}^b$. Go to the node selection step.
6. *Fathoming*: Remove node $i$ from $\mathscr{L}^s$. If $i < k$, set $i = i + 1$ and return to the bounding step. Otherwise, go to the branching step.

One key issue in the branch-and-bound method is to get a good lower bound $B_i$ generated by the bounding step. The better the lower bound, is the more subproblems can be fathomed and the faster the algorithm converges. In the following, we will introduce how to calculate the bound in the bounding step.

#### 2.1.2.2 Bound Calculation

For integer programming problems, continuous relaxation and Lagrange relaxation are two commonly used methods for generating lower bounds.

**Continuous Relaxation** In the continuous relaxation, the integer variables are relaxed into the continuous ones. The continuous relaxation of (2.19) can be expressed as follows:

$$\begin{aligned} &\min f_0(\boldsymbol{x}) \\ \text{s.t. } &f_i(\boldsymbol{x}) \leq 0, i = 1, \ldots, m, \\ &h_j(\boldsymbol{x}) = 0, j = 1, \ldots, p, \\ &\boldsymbol{x} \in conv(\mathscr{X}), \end{aligned} \tag{2.20}$$

where $conv(\mathscr{X})$ is a convex hull of the integer set $\mathscr{X}$. Since the optimal solution of (2.19) is also feasible for (2.20), the solution of (2.20) is the lower bound of (2.19). The continuous relaxed problem can be solved by the aforementioned methods.

**Lagrange Relaxation** As we have introduced in Sect. 2.1.1.1, the dual function yields the lower bound on the optimal value. Define the following Lagrange function for $\boldsymbol{\lambda} \succeq 0$ and $\boldsymbol{\mu} \succeq 0$:

$$L(\boldsymbol{x}, \boldsymbol{\lambda}, \boldsymbol{\mu}) = f_0(\boldsymbol{x}) + \sum_{i=1}^{m} f_i(\boldsymbol{x}) + \sum_{j=1}^{p} h_i(\boldsymbol{x}). \tag{2.21}$$

The Lagrange relaxation problem of (2.19) is given as follows:

$$g(\boldsymbol{\lambda}, \boldsymbol{\mu}) = \min_{\boldsymbol{x} \in \mathscr{X}} L(\boldsymbol{x}, \boldsymbol{\lambda}, \boldsymbol{\mu}). \tag{2.22}$$

According to the results in [4], the Lagrange bound for a convex integer program is at least as good as the bound obtained by the continuous relaxation.

## 2.2 Basics of Game Theory

Game theory can be viewed as a branch of applied mathematics as well as of applied sciences. It has been used in the social sciences, most notably in economics, but has also penetrated into a variety of other disciplines such as political science, biology, computer science, philosophy, as well as wireless and communication networks. In this section, we first give some basic concepts on game theory, and then introduce contract theory which will be used in the following chapters.

### *2.2.1 Basic Concepts*

In this part, we will first introduce the definition of a game and then introduce the concept of *Nash equilibrium*. Finally, we will give some examples of games.

#### 2.2.1.1 Definition of a Game

A game is described by a set of rational *players*, the *strategies* associated with the players, and the *payoffs* for the players. A rational player has his own interest, and therefore will act by choosing an available strategy to achieve his interest. A player is assumed to be able to evaluate exactly or probabilistically the outcome or payoff (usually measured by the utility) of the game which depends not only on his action but also on other players' actions.

Formally, a game in strategic (normal) form is represented by a triple $G = (\mathcal{N}, (\mathcal{S}_i)_{i\in\mathcal{N}}, (u_i)_{i\in\mathcal{N}})$, where

- $\mathcal{N}$ is a finite set of players, i.e., $\mathcal{N} = \{1, \ldots, N\}$;
- $\mathcal{S}_i$ is the set of available strategies for player $i$;
- $u_i : \mathcal{S} \to \mathbb{R}$ is the utility (payoff) function for player $i$, with $\mathcal{S} = \mathcal{S}_1 \times \ldots \times \mathcal{S}_N$.

Given the definition of a strategic game, for player $i$, element $s_i \in \mathcal{S}_i$ is the strategy of player $i$, and $\mathbf{s}_{-i} = [s_j]_{j\in\mathcal{N}, j\neq i}$ denotes the vector of strategies of all players except $i$, and $\mathbf{s} = (s_i, \mathbf{s}_{-i}) \in \mathcal{S}$ is referred to as a *strategy profile*. If the sets of strategies $\mathcal{S}_i$ are finite for all $i \in \mathcal{N}$, the game is called *finite*. It is worthwhile to point out that one user's utility is a function of both this user's and others' strategies.

For a game in strategic form, each player has to select a strategy so as to optimize its utility function. Whenever each player $i \in \mathcal{N}$ selects a strategy $s_i \in \mathcal{S}_i$ in a deterministic manner, i.e., with probability 1, a game is said to be one with *complete information* if all elements of the game are common knowledge. Otherwise, the game is said to be one with *incomplete information*, or an incomplete information game.

#### 2.2.1.2 Nash Equilibrium

Before the definition of the Nash equilibrium, we first introduce the concept of dominating strategies. It can simplify the solution of a game by eliminating some strategies which have no effect on the outcome of the game.

A strategy $s_i \in \mathscr{S}_i$ is said to be *dominant* for player $i$ if

$$u_i(s_i, s_{-i}) \geq u_i(s_i', s_{-i}), \forall s_i' \in \mathscr{S}_i, \forall s_{-i} \in \mathscr{S}_{-i}, \tag{2.23}$$

where $\mathscr{S}_{-i} = \prod_{j \neq i} \mathscr{S}_j$ is the set of all strategy profiles for all players except $i$. Therefore, a dominant strategy is the player's best strategy, i.e., the strategy that yields the highest utility for the player regardless of what strategies the other players choose. Whenever a player has a dominant strategy, a rational player has no incentive to choose any other strategy. Consequently, if each player possesses a dominant strategy, then all players will choose their dominant strategies. A strategy profile $s^* \in \mathscr{S}$ is the *dominant-strategy equilibrium* if element $s_i^*$ of $s^*$ is a dominant strategy of player $i$.

While the dominant-strategy equilibrium is an intuitive solution for a given game, the existence of this equilibrium point is not guaranteed. In fact, there are many games in which no player has a dominant strategy. Therefore, Nash equilibrium is introduced as a solution for a game. Loosely speaking, a Nash equilibrium is a state of a game where no player can improve its utility by changing its strategy, if the other players maintain their current strategies.

A Nash equilibrium of game $G = (\mathscr{N}, (\mathscr{S}_i)_{i \in \mathscr{N}}, (u_i)_{i \in \mathscr{N}})$ is a strategy profile $s^* \in \mathscr{S}$ such that $\forall i \in \mathscr{N}$, we have

$$u_i(s_i^*, s_{-i}^*) \geq u_i(s_i, s_{-i}^*), \forall s_i \in \mathscr{S}_i. \tag{2.24}$$

In other words, a strategy profile is a Nash equilibrium if no player has an incentive to *unilaterally* deviate to another strategy, given that other players' strategies remain fixed.

To explain the Nash equilibrium, we take the famous "Prisoner's Dilemma" as an example. In this example, two suspects are arrested for a crime and placed in isolated rooms. Each one of the suspects has to decide whether or not to confess and implicate the other. The governing rules are the following:

- If none of the prisoners confesses, then each will serve 1 year in jail.
- If both of them confess and implicate each other, they will both go to prison for 3 years.
- If one prisoner confesses and implicates the other while the other one does not confess, the one who has confessed will get free, while the other one will spend 5 years in jail.

In this situation, each prisoner has two strategies: to confess (strategy C) or not to confess (strategy NC). The utility for each prisoner is simply the number of years that will be spent in prison. The matrix representation of this game is given in Table 2.1. The payoffs are negative numbers since we deal with games where the players seek to maximize a utility.

**Table 2.1** Payoff in Prisoner's Dilemma

| | Confess (C) | Not confess (NC) |
|---|---|---|
| Confess (C) | (−3, −3) | (0, −5) |
| Not confess (NC) | (−5, 0) | (−1, −1) |

In this example, each player has a better payoff by confessing, i.e., choosing C, independent of the strategic choice of the other player. Thus, (C,C) is a dominant-strategy equilibrium which yields a payoff vector (2,2). Note that although this point is a solution for the game, the payoffs received are not the best for both players. Also, we can easily find that (C,C) is the only Nash equilibrium of this game.

#### 2.2.1.3 Examples of Game Theory

There exist rich game theoretical approaches, such as Stackelberg game, bargaining game, and auction game. In the following, we will give a brief introduction to these game models as examples. If the readers are interested in other game models, please refer to [5].

1. **Stackelberg Game**: In many non-cooperative games, a hierarchy among the players might exist whereby one or more of the players declare and announce their strategies before the other players select their strategies. In such a hierarchical framework, the declaring players can be in a position to enforce their own strategies upon the other players. Therefore, in these games, the player who holds the strong position and that can impose its own strategy upon the others is called the *leader* while the players who react to the leader's declared strategy are called *followers*. The games between leaders and followers are called Stackelberg game.
2. **Bargaining Game**: In economics, many problems involve a number of entities that are interested in reaching an agreement over a trade or the sharing of a resource but have a conflicting interest on how to reach this agreement and on the terms of the agreement. In this context, a bargaining situation is defined as a situation in which two (or more) players can mutually benefit from reaching a certain agreement but have conflicting interests on the terms of the agreement. Certainly, in a bargaining situation, no agreement can be imposed on any player without the player's approval. Bargaining game theory is established to deal with studying and analyzing bargaining situations in a variety of problems.
3. **Auction Game**: Auction theory is an applied branch of game theory that deals with how people act in auction markets, and it studies the game-theoretic properties of auction markets. There are many possible designs (or sets of rules) for an auction, and typical issues studied by auction theorists include the efficiency of a given auction design, optimal and equilibrium bidding strategies, and revenue comparison. Auction theory is also used as a tool to inform the design of real-world auctions, e.g., auctions for the privatization of public-sector companies or the sale of licenses for use of the electromagnetic spectrum.

## 2.2.2 *Contract Theory*

Contract theory is widely used in real-world economics with asymmetric information to design contracts between employer(s) and employee(s) by introducing cooperation [6]. The information asymmetry usually refers to the fact that the employer(s) does not know exactly the characteristics of the employee(s). By using contract theory-based models, the employer(s) can overcome this asymmetric information and efficiently incentivize its employee(s) by offering a contract which includes a given performance and a corresponding reward.

Due to the properties such as inducing cooperation and dealing with asymmetric information, we envision that there is a great potential to utilize concepts from contract theory to ensure cooperation and assist in the design of incentive mechanisms in wireless networks. In wireless networks, the employer(s) and employee(s) can be of different roles depending on the scenario under consideration. Well-designed contracts provide incentives for the contracting parties to exploit the prospective gains from cooperation.

In the following, we first introduce two types of contract problems, and then elaborate on the models and reward design.

### 2.2.2.1 Classification

Contract theory allows studying the interaction between employer(s) and employee(s). The performance of employees tends to be better when they work harder, and the probability of a bad performance will be lower if employees place more dedication or focus on the work. By contrast, if an employee's compensation is independent of its performance, the employee will be less likely to put efforts into the work. The design of incentive mechanism plays an important role in addressing the problem of employee incentives.

In contract theory, the solution we need to obtain is a menu of contract for the employee, and the object is to maximize the employer's payoff or utility. In most cases, the problem is usually formulated as maximizing an objective function which represents the employer's payoff, subject to the **Incentive Compatibility** (IC) constraint that the employee's expected payoff is maximized when signing the contract and the **Individual Rationality** (IR) constraint that the employee's payoff under this contract is larger than or equal to its reservation payoff when not participating.

Typically, the contract problems can be classified into the following two types [7]:

**Adverse Selection** The adverse selection problem and the information about some relevant characteristics of the employees, such as their distaste for certain tasks and their level of competence/productivity, are hidden from the employer. One of the most common problems in adverse selection is the *screening problem*, in which the contract is offered by the uninformed party, i.e., the employer. The uninformed party typically responds to adverse selection by the revelation principle which forces the informed party to select contract that fits its true status.

In the screening problem, the employer makes the contract offer and tries to screen the information hold by the employee. Based on the revelation principle, the employer can offer multiple employment contracts $(q, r)$ destined to different skill-level employees, where $q$ is the employee's outcome wanted by the employer, and $r$ is the reward paid to the employee by the employer if the given target is achieved. The outcome can be a required performance, or some other outcomes that the employer wants from the employee.

Most of the adverse selection models have the following system model. Assume that there are $n$ different types, $\theta_i, i \in \mathcal{N} = \{1, \ldots, n\}$, of employees, where the type represents their level of capability or competence. There exists an information asymmetry that the employer does not know the exact type of employee, but only the probability $\lambda_i$ of facing a type of employee $\theta_i$.

The employer has the expected utility function

$$U = \sum_{i=1}^{n} \lambda_i (q_i - c r_i), \tag{2.25}$$

where $q_i$ is the employer's received outcome from the employee, $r_i$ is the reward that the employer has to pay, and $c$ is a constant. The employee has the utility function

$$V = \theta_i v(r_i) - q_i, \tag{2.26}$$

where $v(r_i)$ is the employee's evaluation toward the reward received from the employer and $q_i$ is the cost of the outcome.

Therefore, aiming to maximize the employer's utility, the problem can be written by

$$\begin{aligned} &\max_{(q,r)} U, \\ s.t.\ &\text{(IR)}\ \theta_i v(r_i) - q_i \geq 0, \\ &\text{(IC)}\ \theta_i v(r_i) - q_i \geq \theta_i v(r_j) - q_j, i, j \in \mathcal{N}, i \neq j. \end{aligned} \tag{2.27}$$

The IR constraints mean the contract needs to guarantee a nonnegative utility for all types of employees. The IC constraint guarantees that the employee can only receive the highest utility when selecting the contract designed for their own types. The model can also be extended to the continuous type to fit the more general cases.

**Moral Hazard** The problem of moral hazard refers to situations where the employee's actions are hidden from the employer: whether they work or not, how hard they work, how careful they are. In contrast to adverse selection, the informational asymmetries in moral hazard arise after the contract has been signed. In moral hazard, the contract is a menu of action–reward bundle $(a, r)$, where $a$ is the action or effort exerted by the employee after being hired and $r$ is the reward paid to the employee by the employer.

The basic model of the moral hazard problem is as below. The employer offers a compensation package $r$ to an employee, which is a combination of a fixed salary $t$ and a performance-related bonus $s$. The employee's performance $q$ can be defined according to the application. During work time, the employee's effort can be regarded as taking an action $a$, while there is asymmetric information that the effort a is hidden from the employer who can only observe the performance $q$. Due to some measurement errors, the performance $q$ is slightly different from the actual effort exerted by the employee. Therefore, the performance of the employee is a noisy signal of its effort. Thus, we assume that the performance $q$ to be normally distributed with mean $a$ and variance $\sigma_q^2$:

$$q = a + \epsilon_q, \tag{2.28}$$

where $\epsilon_q \sim N(\mu_q, \sigma_q^2)$, and $\mu_q$ is the mean. One simple form for the bonus is the linear form. By restricting the compensation package offered by the employer in the linear form, the compensation package $r$ that the employee receives from the employer can be written as

$$r = t + sq, \tag{2.29}$$

where $t$ denotes the fixed compensation salary independent of performance, and $s$ is the fraction of reward related to the employee's performance $q$.

The employee is usually assumed to have a constant attitude toward risk as its income increases. Thus, employee's utility is represented by a negative exponential utility form:

$$u(w, a) = -\exp(\eta(r - \phi(a))), \tag{2.30}$$

where $\eta > 0$ is the employee's coefficient of absolute risk aversion. A larger value of $\eta$ means a less incentive for the employee to implement an effort. $\phi(a)$ is the cost to provide the effort a for the employer. The cost function can be assumed to be quadratic, i.e.,

$$\phi(a) = \frac{1}{2}ca^2. \tag{2.31}$$

The utility of the employer is the evaluation of the outcome $q$ minus the compensation package $r$ to the employee. That is,

$$U(r, a) = E(q - r), \tag{2.32}$$

where $E(.)$ is the evaluation function follows $E(0) = 0$, $E'(.) > 0$, and $E''(.) \geq 0$. If the employer is assumed to be risk neutral, i.e., $E''(.) = 0$. Thus, the utility can be simplified as

$$U(r, a) = q - r = (1 - s)q - t. \tag{2.33}$$

Therefore, the problem of moral hazard is usually formulated as follows:

$$\begin{aligned} &\max_{a,t,s} U(r,a), \\ s.t.\ &(\text{IC})\ \ a^* \in \arg\max_{a} U(r,a), \\ &(\text{IR})\ \ U(r,a) \geq U(r). \end{aligned} \tag{2.34}$$

The IR constraint indicates that the contract must ensure the employee receives a higher utility than when it is not participated in. The IC constraint guarantees that the employee can maximize its own utility when selecting the right amount of effort.

#### 2.2.2.2 Models and Reward Design

**Bilateral or Multilateral** Bilateral contracting is the basic one-to-one contracting model, in which there are one employer and one employee trading with each other for goods or services. However, in the multilateral case, it is usually a one-to-many contracting scenario, in which one employer trades with multiple employees. Despite the increased number of participants in the multilateral contracting than in the bilateral one, the interactions among the employees/buyers, such as competition and cooperation, make the multilateral contracting model more complex and show the potential of solving more sophisticated problems.

Next, we are going to talk about how to design reward in bilateral and multilateral contracting scenarios.

1. *Contract with Single Employee*: When the employer signs a contract with a single employee, we can design the reward by considering only the single employee's absolute performance instead of the others. However, even though there is no other employee to compete with the employee, the relative performance-related reward can still be applied. One common form of the relative performance-related reward for a single employee is to set up a specific threshold and a reward of the targeted performance. If the employee's absolute performance can achieve the given threshold, a fixed reward will be given to the employee. Otherwise, the employee cannot receive the reward. In fact, we can regard it as the employee competing with the threshold.
2. *Contract with Multi-employee*: When the employer designs the contract toward multiple employees, the absolute performance-related reward and is a widely adopted method in real economics. Furthermore, there are some other forms of absolute performance-related rewards. One commonly used method is to group employees first and then reward employees by their aggregated performance in each group. However, there is a shortcoming with this incentive mechanism, i.e., there is a risk of free riding of some employees on the other employees' efforts. Usually, the absolute performance-related reward design is more commonly seen in contracting with multi-employee. The employees can compete with each other as in a tournament and have the incentives for higher rewards by performing better.

**One-Dimension or Multi-Dimension** Only one characteristic or task is considered in the one-dimensional contracting model. For example, the employer evaluates only one capability of the employee in the one-dimensional adverse selection model, and only one task is assigned by the employer to the employee in the one-dimensional moral hazard model. In contrast, the employer evaluates multi-dimensional characteristics of the employee or assigns multiple tasks to the employee in the multi-dimensional contracting scenario. For example, action $a$ in the one-dimensional moral hazard model can be extended to $\boldsymbol{a} = (a_1, \ldots, a_n), n \geq 2$. Meanwhile, the observed performance becomes $\boldsymbol{q} = (q_1, \ldots, q_n)$, as well as the bonus $\boldsymbol{s} = (s_1, \ldots, s_n)$. As the extension of one-dimensional contracting, multi-dimensional contracting model can also be analyzed by adapting the similar methods for one-dimensional ones.

One-dimensional model becomes inefficient when employees are required to have multiple capabilities or supposed to work on several tasks. First, the employee's action set becomes richer than that in the one-dimensional model. Second, there is a risk that one-dimensional reward will induce employees to overwhelmingly focus on the part that will be rewarded and to neglect the other components. Taking Yelp for an example, which is a popular mobile crowdsourcing app in North America used to locate and review restaurants/bars, Yelp users who act as employees do not only make location-based check-ins, upload photographs, and write reviews of the restaurants and bars, but they are also encouraged to invite new friends to sign up. If Yelp only rewards users on the number of reviews, the quality of a review such as length, correctness, and objectiveness will not be considered. Given different aspects of capability or multiple tasks to evaluate, by assigning different weights of rewards in multiple dimensions, the employer can drive employee's incentive on perusing certain capabilities or tasks, which can affect the employer's utility, in return.

Thus, in certain scenarios, one-dimensional reward needs to be modified into multi-dimensional ones, so that the employer can drive employee's incentives by assigning different reward weights on different tasks. Regardless of the dimension that the reward design chooses, a qualified mechanism must reward employee's effort in a comprehensive way. On the one hand, for simple cases where a one-dimensional reward is sufficient to drive employee's motivation, a multi dimensional reward mechanism costs extra effort and resource to design. On the other hand, for complicated tasks, the reward design must be adjusted to multi-dimension, so that employee's incentive can be well maintained and driven. In reward design, there is a trade-off between completion and efficiency, and thus, we should model the dimension of the reward according to the actual scenario under consideration.

It is also worthwhile to point out that in auction theory, there is one seller with an item to sell and multiple bidders with reservation prices competing for it. Meanwhile, in the multilateral adverse selection, there are one seller and multiple buyers with their own private information which is the same case as the bidder's reservation prices during the auction. Therefore, auction theory is a special case of the multilateral adverse selection contracting problem in contract theory.

## 2.3 Related Machine Learning Technologies

Machine learning is a part of artificial intelligence. To be intelligent, a system that is in a changing environment should have the ability to learn. If the system can learn and adapt to such changes, the system designer need not foresee and provide solutions for all possible situations. In other words, machine learning enables computers to optimize a performance criterion using example data or past experience without being explicitly programmed.

In the following, we will first introduce some kinds of classical machine learning technologies, and then present deep learning and reinforcement learning technologies to tackle the scenarios where the classical ones are not efficient.

### 2.3.1 *Classical Machine Learning*

A machine learning algorithm is an algorithm that is able to learn from data. Mitchell provides the definition [8]: A computer program is said to learn from experience $E$ with respect to some class of tasks $T$ and performance measure $P$, if its performance at tasks in $T$, as measured by $P$, improves with experience $E$.

**Task** $T$ Machine learning allows us to tackle tasks that are too difficult to solve with fixed programs written and designed by human beings. Many kinds of tasks can be solved with machine learning. Some of the most common machine learning tasks include the following:

- Classification: In this type of task, the computer program is asked to specify which of $k$ categories some input belongs to. To solve this task, the learning algorithm produces a function $f : \mathbb{R}^n \rightarrow \{1, \ldots, k\}$. When $y = f(\boldsymbol{x})$, the function assigns an input $\boldsymbol{x}$ to a category $y$.
- Regression: In this type of task, the computer program is asked to predict a value given some input. To solve this task, the learning algorithm is asked to output a function $f : \mathbb{R}^n \rightarrow \mathbb{R}$. This type of task is similar to classification, except that the format of output is different.
- Denoising: In this type of task, the machine learning algorithm is given in input a corrupted example $\tilde{\boldsymbol{x}} \in \mathbb{R}^n$ obtained by an unknown corruption process from a clean example $\boldsymbol{x} \in \mathbb{R}^n$. The algorithm must predict the clean example $\boldsymbol{x}$ from its corrupted version $\tilde{\boldsymbol{x}}$.

**Performance Measure** $P$ In order to evaluate the abilities of a machine learning algorithm, we must design a quantitative measure of its performance. Usually we are interested in how well the machine learning algorithm performs on data that it has not seen before, since this determines how well it will work when deployed in the real world. We therefore evaluate these performance measures using a test set of data that is separate from the data used for training the machine learning system. The choice of performance measure may seem straightforward, but it is often difficult to choose a performance measure that corresponds well to the desired behavior of the system.

**Experience** $E$ Machine learning algorithms can be broadly categorized as *unsupervised* or *supervised* by what kind of experience they are allowed to have during the learning process. Unsupervised learning algorithms experience a dataset containing many features, then learn useful properties of the structure of this dataset. Supervised learning algorithms experience a dataset containing features, but each example is also associated with a label or target.

Roughly speaking, unsupervised learning algorithms observe several examples of a random vector $\boldsymbol{x}$, and attempt to implicitly or explicitly learn the probability distribution $p(\boldsymbol{x})$, or some interesting properties of that distribution, while supervised learning algorithms observe several examples of a random vector $\boldsymbol{x}$ and an associated value $y$, and predict $y$ from $\boldsymbol{x}$ by estimating $P(y|\boldsymbol{x})$. The supervised learning originates from the view of the target $y$ being provided by an instructor who shows the machine learning system what to do. In unsupervised learning, there is no instructor, and the algorithm must learn to make sense of the data without this guide.

Though unsupervised learning and supervised learning are not distinct concepts, they do help us roughly categorize some things we do with machine learning algorithms. In the following, we will elaborate on supervised and unsupervised learning. Finally, we will introduce how to design a machine learning algorithm.

#### 2.3.1.1 Supervised Learning

Supervised learning are learning algorithms that learn to associate some input with some output, given a training set of examples of inputs $\boldsymbol{x}$ and outputs $y$. In many cases the outputs $y$ must be provided by a human "supervisor." One of the most influential approaches to supervised learning is the support vector machine (SVM). For example, in the classification problem, if we have two classes, i.e., positive and negative classes, SVM predicts that $\boldsymbol{x}$ belongs to the positive class when $\boldsymbol{w}^T\boldsymbol{x} + b$ is positive. Likewise, it predicts that it belongs the negative class when $\boldsymbol{w}^T\boldsymbol{x} + b$ is negative.

One key innovation associated with SVM is the kernel trick. The kernel trick is that many machine learning algorithms can be written exclusively in terms of dot products between examples. For example, the linear function used by the SVM can be written by

$$\boldsymbol{w}^T\boldsymbol{x} + b = b + \sum_{i=1}^{n} \alpha_i \boldsymbol{x}^T\boldsymbol{x}^i, \tag{2.35}$$

where $\boldsymbol{x}^i$ is a training example and $\boldsymbol{\alpha}$ is a vector of coefficients. Replace $\boldsymbol{x}$ by the output of a given feature function $\phi(\boldsymbol{x})$ and the dot product can be replaced with a function $k(\boldsymbol{x}, \boldsymbol{x}^i) = \phi(\boldsymbol{x})^T\phi(\boldsymbol{x}^i)$, which is called a *kernel*. After adopting the kernel function, we can make predictions using the function

$$f(\boldsymbol{x}) = b + \sum_{i=1}^{n} \alpha_i k(\boldsymbol{x}, \boldsymbol{x}^i). \tag{2.36}$$

This function is nonlinear with respect to $\boldsymbol{x}$, but the relationship between $\phi(\boldsymbol{x})$ and $f(\boldsymbol{x})$ is linear. The kernel-based function is exactly equivalent to preprocessing the data by applying $\phi(\boldsymbol{x})$ to all inputs, then learning a linear model in the transformed space. The kernel trick is powerful for two reasons. First, it allows us to learn models that are nonlinear as a function of $\boldsymbol{x}$ using convex optimization techniques that are guaranteed to converge efficiently. This is possible because we consider $\phi$ fixed and optimize only $\boldsymbol{\alpha}$. Second, the kernel function $k(.)$ often admits an implementation that is significantly more computational efficient than naively constructing two $\phi(\boldsymbol{x})$ vectors and explicitly taking their dot product.

The most commonly used kernel is the Gaussian kernel

$$k(\boldsymbol{u}, \boldsymbol{v}) = \mathcal{N}(\boldsymbol{u} - \boldsymbol{v}, 0, \sigma^2 \boldsymbol{I}), \tag{2.37}$$

where $\mathcal{N}(\boldsymbol{x}, \boldsymbol{\mu}, \boldsymbol{\Sigma})$ is the standard normal density. The Gaussian kernel corresponds to a dot product in an infinite-dimensional space. We can regard the Gaussian kernel as performing a kind of template matching [9]. A training example $\boldsymbol{x}$ associated with training label $y$ becomes a template for class $y$. When a test point $\boldsymbol{x}_0$ is near $\boldsymbol{x}$ according to Euclidean distance, the Gaussian kernel has a large response, indicating that $\boldsymbol{x}_0$ is very similar to the $\boldsymbol{x}$ template. The model then puts a large weight on the associated training label $y$. Overall, the prediction will combine many such training labels weighted by the similarity of the corresponding training examples.

#### 2.3.1.2 Unsupervised Learning

Unsupervised algorithms are those that experience only "features" but not a supervision signal. Informally, unsupervised learning refers to most attempts to extract information from a dataset that do not require human to annotate examples. A classic unsupervised learning task is to find a representation that preserves as much information about input $\boldsymbol{x}$ as possible while obeying some penalty or constraint aimed at keeping the representation simpler or more accessible than $\boldsymbol{x}$ itself. In this part, we introduce a typical unsupervised learning algorithm, i.e., $k$-means clustering.

The $k$-means clustering algorithm divides the training set into $k$ different clusters of examples that are near each other. We can thus think of the algorithm as providing a $k$-dimensional vector $\boldsymbol{h}$ representing an input $\boldsymbol{x}$. If $\boldsymbol{x}$ belongs to cluster $i$, then $h_i = 1$ and all other entries of $\boldsymbol{h}$ are zero.

The $k$-means algorithm works by initializing $k$ different centroids $\{\boldsymbol{\mu}^{(1)}, \ldots, \boldsymbol{\mu}^{(k)}\}$ to different values, then alternating between two different steps until convergence.

- In one step, each training example is assigned to cluster $i$, where $i$ is the index of the nearest centroid $\boldsymbol{\mu}^{(i)}$.
- In the other step, each centroid $\boldsymbol{\mu}^{(i)}$ is updated to the mean of all training examples $\boldsymbol{x}^{(j)}$ assigned to cluster $i$.

One difficulty of clustering is that there is no single criterion that measures how well a clustering result of the data corresponds to the real world. We can measure properties of the clustering such as the average Euclidean distance from a cluster centroid to the members of the cluster. This allows us to tell how well we are able to reconstruct the training data from the cluster assignments. We do not know how well the cluster assignments correspond to properties of the real world. Moreover, there may be many different clusterings that all correspond well to some property of the real world. We may hope to find a clustering result that relates to one feature but obtain a different, equally valid clustering that is not relevant to our task.

#### 2.3.1.3 Machine Learning Algorithm Design

Nearly all machine learning algorithms can be described as particular instances of a simple recipe: a dataset, a cost function, an optimization procedure, and a model [9].

For example, the linear regression algorithm combines a dataset consisting of $\boldsymbol{X}$ and $\boldsymbol{y}$, the cost function

$$J(\boldsymbol{w}, b) = \mathbb{E}_{\boldsymbol{x},y} \log p_{model}(y|\boldsymbol{x}), \tag{2.38}$$

the model specification $p_{model}(y|\boldsymbol{x}) = \mathcal{N}(y, \boldsymbol{x}^T \boldsymbol{w} + b, 1)$, and the optimization algorithm in most cases is to set the gradient of the cost as zero.

By realizing that we can replace any of these components mostly independently from the others, we can obtain a very wide variety of algorithms.

The cost function typically includes at least one term that causes the learning process to perform statistical estimation. The most common cost function is the negative log-likelihood, so that minimizing the cost function causes maximum likelihood estimation.

If we change the model to be nonlinear, then most cost functions can no longer be optimized in closed form. This requires us to choose an iterative numerical optimization procedure, such as gradient descent. In some cases, the cost function may be a function that we cannot actually evaluate, for computational reasons. In these cases, we can still approximately minimize it using iterative numerical optimization so long as we have some way of approximating its gradients.

### 2.3.2 Deep Learning

The classical machine learning algorithms introduced in the last subsection work very well on a wide variety of important problems. However, they have not succeeded in solving the central problems in artificial intelligence, such as recognizing speech or recognizing objects due to the following reasons [9]:

1. **The Curse of Dimensionality**: Many machine learning problems become exceedingly difficult when the number of dimensions in the data is high. This phenomenon is known as the curse of dimensionality. Of particular concern is that the number of possible distinct configurations of a set of variables increases exponentially as the number of variables increases.
2. **Local Constancy and Smoothness Regularization**: In order to generalize well, machine learning algorithms need to be guided by prior beliefs about what kind of function they should learn. Previously, we have seen these priors incorporated as explicit beliefs in the form of probability distributions over parameters of the model. Among the most widely used of these implicit "priors" is the smoothness prior or local constancy prior. This prior states that the function we learn should not change very much within a small region. Many simpler algorithms rely exclusively on this prior to generalize well, and as a result they fail to scale to the statistical challenges involved in solving artificial intelligent level tasks.
3. **Manifold Learning**: An important concept underlying many ideas in machine learning is a manifold. A manifold is a connected region. Mathematically, it is a set of points, associated with a neighborhood around each point. From any given point, the manifold locally appears to be a Euclidean space. In everyday life, we experience the surface of the world as a 2D plane, but it is in fact a spherical manifold in 3D space. Although there is a formal mathematical meaning to the term "manifold" in machine learning, it tends to be used more loosely to designate a connected set of points that can be approximated well by considering only a small number of degrees of freedom, embedded in a higher-dimensional space. Each dimension corresponds to a local direction of variation. In the context of machine learning, we allow the dimensionality of the manifold to vary from one point to another. This often happens when a manifold intersects itself.

These challenges motivate the development of deep learning. In the following, we will introduce the basics of neural networks and the back-propagation algorithms in detail.

#### 2.3.2.1 Basics of Neural Networks

Neural networks are the quintessential deep learning models. The goal of a neural network is to approximate some function $f^*$. For example, for a classifier, $y = f^*(\boldsymbol{x})$ maps an input $\boldsymbol{x}$ to a category $y$. A neural network defines a mapping $y = f(\boldsymbol{x}, \theta)$ and learns the value of the parameters $\theta$ that result in the best function approximation.

Neural networks are called networks because they compose many different functions. The model is associated with a directed acyclic graph describing how the functions are composed together. For example, we might have three functions $f^{(1)}$, $f^{(2)}$, and $f^{(3)}$ connected in a chain to form $f(\boldsymbol{x}) = f^{(3)}(f^{(2)}(f^{(1)}(\boldsymbol{x})))$. These chain structures are the most commonly used in neural networks. In this case, $f^{(1)}$ is called the first layer of the network, $f^{(2)}$ is called the second layer, and

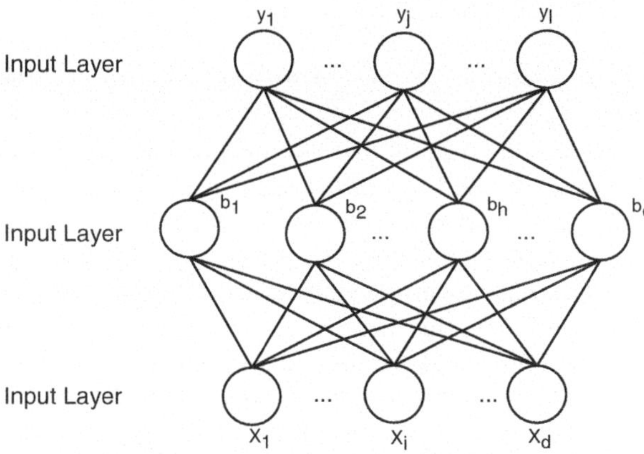


**Fig. 2.1** Example of a neural network

so on. The overall length of the chain gives the depth of the model. It is from this terminology that the name "deep learning" arises. An example of the neural network is given in Fig. 2.1. The final layer of a neural network is called the *output layer*.

During neural network training, we drive $f(\boldsymbol{x})$ to match $f^*(\boldsymbol{x})$. The training data provides us with noisy, approximate examples of $f^*(\boldsymbol{x})$ evaluated at different training points. Each example $\boldsymbol{x}$ is accompanied by a label $y = f^*(\boldsymbol{x})$. The training examples specify directly what the output layer must do at each point $\boldsymbol{x}$. It must produce a value that is close to $y$. The behavior of the other layers is not directly specified by the training data. The learning algorithm must decide how to use those layers to produce the desired output, but the training data does not say what each individual layer should do. Instead, the learning algorithm must decide how to use these layers to best implement an approximation of $f^*$. Because the training data does not show the desired output for each of these layers, these layers are called *hidden layers*.

Designing and training a neural network is not much different from training any other machine learning model with gradient descent. The largest difference lies in that the nonlinearity of a neural network causes most interesting loss functions to become non-convex. This means that neural networks are usually trained by iterative and gradient-based optimizers that merely drive the cost function to a very low value. For neural networks, it is important to initialize all weights to small random values. The biases may be initialized to zero or to small positive values. In what follows, we will introduce how to train a neural network, which includes cost functions, output units, and hidden units.

1. **Cost Functions**: An important aspect of the design of a deep neural network is the choice of the cost function. Fortunately, the cost functions for neural networks are more or less the same as those for other parametric models. In most cases, our parametric model defines a distribution $P(y|\boldsymbol{x}, \boldsymbol{\theta})$ and we simply use the principle of maximum likelihood. This means we use the cross-entropy between the training data and the model's predictions as the cost function.

This cost function is given by

$$J(\boldsymbol{\theta}) = \mathbb{E}_{\boldsymbol{x},y} \log p_{model}(y|\boldsymbol{x}), \tag{2.39}$$

The specific form of the cost function changes from model to model, depending on the specific form of $\log p_{model}(y|\boldsymbol{x})$. One important observation throughout neural network design is that the gradient of the cost function must be large and predictable enough to serve as a good guide for the learning algorithm. Functions that saturate undermine this objective because they make the gradient very small. In many cases this happens because the activation functions used to produce the output of the hidden units or the output units saturate. The negative log-likelihood helps to avoid this problem for many models.

One unusual property of the cross-entropy cost used to perform maximum likelihood estimation is that it usually does not have a minimum value when applied to the models commonly used in practice. For discrete output variables, most models are parameterized in such a way that they cannot represent a probability of zero or one, but can come arbitrarily close to doing so.

2. **Output Units**: The choice of cost function is tightly coupled with the choice of output unit. Most of the time, we simply use the cross-entropy between the data distribution and the model distribution. The choice of how to represent the output then determines the form of the cross-entropy function. Any kind of neural network unit that may be used as an output can also be used as a hidden unit. Here, we focus on the use of these units as outputs of the model, but in principle they can be used internally as well. In this part, we suppose that the neural network provides a set of hidden features defined by $\boldsymbol{h} = f(\boldsymbol{x}, \boldsymbol{\theta})$. The role of the output layer is then to provide some additional transformation from the features to complete the task that the network must perform.

   One simple kind of output unit is an output unit based on an affine transformation. These are often just called linear units. Given features $\boldsymbol{h}$, a layer of linear output units produces a vector $\hat{\boldsymbol{y}} = \boldsymbol{W}^T\boldsymbol{h} + \boldsymbol{b}$. Linear output layers are often used to produce the mean of a conditional

$$p(\boldsymbol{y}|\boldsymbol{x}) = \mathcal{N}(\boldsymbol{y}, \hat{\boldsymbol{y}}, \boldsymbol{I}). \tag{2.40}$$

   Maximizing the log-likelihood is then equivalent to minimizing the mean squared error. The maximum likelihood framework makes it straightforward to learn the covariance of the Gaussian too, or to make the covariance of the Gaussian be a function of the input. However, the covariance must be constrained to be a positive definite matrix for all inputs. It is difficult to satisfy such constraints with a linear output layer, so typically other output units are used to parameterize the covariance. Since linear units do not saturate, they pose little difficulty for gradient-based optimization algorithms and may be used with a wide variety of optimization algorithms.

3. **Hidden Units**: The design of hidden units is an extremely active area of research and does not yet have many definitive guiding theoretical principles. Rectified linear units are an excellent default choice of hidden unit. Many other types of hidden units are available. It can be difficult to determine when to use which kind (though rectified linear units are usually an acceptable choice). Here, we describe some basic observations on the hidden units.

The design process consists of trial and error, intuiting that a kind of hidden unit may work well, and then training a network with that kind of hidden unit and evaluating its performance on a validation set. Some of the hidden units included in this list are not actually differentiable at all input points, which implies that a gradient-based learning algorithm cannot be used. In practice, gradient descent still performs well enough for these models to be used for machine learning tasks. This is in part because neural network training algorithms do not usually achieve a local minimum of the cost function, but instead merely reduce its value significantly. Because we do not expect training to actually reach a point where the gradient is 0, it is acceptable for the minima of the cost function to correspond to points with undefined gradient. An important point in practice is that one can safely disregard the non-differentiability of the hidden unit activation functions.

Rectified linear units use the activation function $g(z) = \max\{0, z\}$. Rectified linear units are widely used because they are so similar to linear units. The only difference between a linear unit and a rectified linear unit is that a rectified linear unit outputs zero across half its domain. This makes the derivatives through a rectified linear unit remain large whenever the unit is active. The gradients are not only large but also consistent. The second derivative of the rectifying operation is 0 almost everywhere, and the derivative of the rectifying operation is 1 everywhere that the unit is active. This means that the gradient direction is far more useful for learning than it would be with activation functions that introduce second-order effects.

Rectified linear units are typically used on top of an affine transformation:

$$\boldsymbol{h} = g(\boldsymbol{W}^T \boldsymbol{x} + \boldsymbol{b}). \tag{2.41}$$

When initializing the parameters of the affine transformation, it can be a good practice to set all elements of $\boldsymbol{b}$ to a small and positive value. This makes it very likely that the rectified linear units will be initially active for most inputs in the training set and allow the derivatives to pass through.

#### 2.3.2.2 Back-Propagation Algorithm

When we use a neural network to produce an output $\hat{\boldsymbol{y}}$ with input $\boldsymbol{x}$, information flows forward through the network. The inputs $\boldsymbol{x}$ provide the initial information that then propogate up to the hidden units at each layer and outputs $\hat{\boldsymbol{y}}$. This is called forward propagation. During training, forward propagation can continue onward until it produces a scalar cost $J(\boldsymbol{\theta})$. The back-propagation algorithm [10] allows the information from the cost to then flow backwards through the network, in order to compute the gradient. Computing an analytical expression for the gradient is straightforward, but numerically evaluating such an expression can be computationally expensive. The back-propagation algorithm does so using a simple and inexpensive procedure. Actually, back-propagation refers only to the method for computing the gradient, while another algorithm, such as stochastic gradient descent, is used to perform learning using this gradient.

To help describe the back-propagation algorithm more precisely, we first introduce the computational graph language. Many ways of formalizing computation as graphs are possible. Here, we use each node in the graph to present a variable, which can be a scalar, vector, matrix, or tensor. To formalize our graphs, we also need to introduce the concept of an operation. An operation is a simple function of one or more variables. Our graph language is accompanied by a set of allowable operations. Without loss of generality, we define an operation to return only a single output variable. For example, if a variable $y$ is computed by applying an operation to a variable $x$, then they will be connected by a directed edge. Examples of computational graphs are given in Fig. 2.2. In the following, we will elaborate on the back-propagation algorithm that specifies the actual gradient computation directly.

First, consider a computational graph describing how to compute a single scalar $u^{(n)}$. This scalar is the variable whose gradient we want to obtain, with respect to the $n_i$ input nodes $\{u^{(1)}, \ldots, u^{(n_i)}\}$, that is, we want to compute $\frac{\partial u^{(n)}}{\partial u^{(i)}}$, for all $i \in \{1, \ldots, n_i\}$. We will assume that the nodes of the graph have been ordered in such a way that we can compute their output one after the other, starting at $u(n_{i+1})$

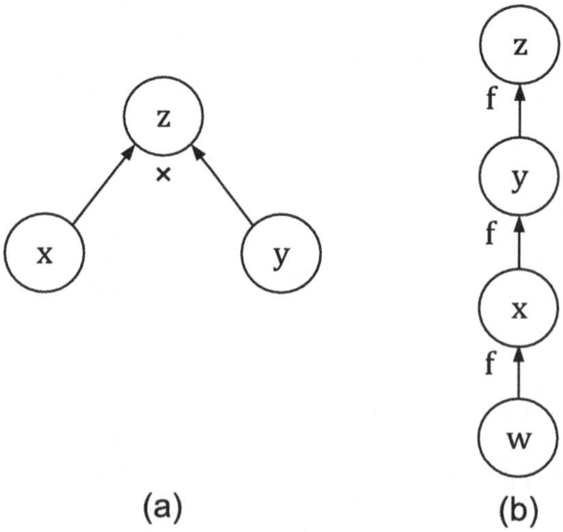


**Fig. 2.2** Examples of computational graphs: (**a**) The graph computes $z = x \times y$. (**b**) We use the same function $f : \mathbb{R} \to \mathbb{R}$ to obtain a chain: $x = f(w)$, $y = f(x)$, and $z = f(y)$

and going up to $u^{(n)}$. We define that each node $u^{(i)}$ is associated with an operation $f^{(i)}$ and it can be computed by

$$u^{(i)} = f^{i}(\mathcal{U}^{(i)}), \tag{2.42}$$

where $\mathcal{U}^{(i)}$ is the set of all nodes that are parents of $u^{(i)}$.

In order to perform the back-propagation, we can construct a computational graph $B$ that depends on the forward propagation network $G$. Computation in $B$ proceeds in exactly the reverse of the order of computation in $G$, and each node of $B$ computes the derivative $\frac{\partial u^{(n)}}{\partial u^{(i)}}$ associated with the forward graph node $u^{(i)}$, and this can be done by

$$\frac{\partial u^{(n)}}{\partial u^{(j)}} = \sum_{i:j\in\mathcal{U}^{(i)}} \frac{\partial u^{(n)}}{\partial u^{(i)}} \frac{\partial u^{(i)}}{\partial u^{(j)}}. \tag{2.43}$$

The graph $B$ contains exactly one edge for each edge from node $u^{(j)}$ to node $u^{(i)}$ of $G$. The edge from $u^{(j)}$ to $u^{(i)}$ is associated with the computation of $\frac{\partial u^{(i)}}{\partial u^{(j)}}$. In addition, a dot product is performed for each node, between the gradient already computed with respect to nodes $u^{(i)}$ that are children of $u^{(j)}$. To summarize, the amount of computation required for performing the back-propagation scales linearly with the number of edges in $G$, where the computation for each edge corresponds to computing a partial derivative as well as performing one multiplication and one addition.

For example, assume that $B$ is given in Fig. 2.2b. To compute $\frac{\partial z}{\partial w}$, we have the following equation according to (2.43):

$$\frac{\partial z}{\partial w} = \frac{\partial z}{\partial y}\frac{\partial y}{\partial x}\frac{\partial x}{\partial w} = f'(y)f'(x)f'(w) \tag{2.44}$$

$$= f'(f(f(w)))f'(f(w))f'(w). \tag{2.45}$$

Equation (2.44) suggests that we can compute the value of $f(w)$ only once and store it in the variable $x$. An alternative approach is suggested by Eq. (2.45), where the subexpression $f(w)$ appears more than once. When the memory required to store the value of these expressions is sufficient, the back-propagation algorithm is preferred because of it can reduce the running time. However, Eq. (2.45) is useful when memory is limited.

The back-propagation algorithm is designed to reduce the number of common subexpressions without regard to memory. Specifically, it performs on the order of one Jacobian product per node in the graph. The back-propagation algorithm visits each edge from node $u^{(j)}$ to node $u^{(i)}$ of the graph exactly once in order to obtain the associated partial derivative $\frac{\partial u^{(i)}}{\partial u^{(j)}}$. Back-propagation thus avoids the exponential explosion in repeated subexpressions.

## 2.3.3 *Reinforcement Learning*

In addition to the aforementioned machine learning algorithms which require a fixed dataset, some machine learning algorithms do not just experience a fixed dataset. For example, reinforcement learning algorithms interact with an environment, so there is a feedback loop between the learning system and its experiences.

A typical reinforcement learning (RL) problem [11] refers to the learning of the agent(s) to react for the environment so as to maximize some numerical value which represents a long-term objective. A typical setting where reinforcement learning operates is shown in Fig. 2.3. An agent receives the environment's state and a reward associated with the last state transition. It then calculates an action. In response, the environment makes a transition to a new state and the cycle is repeated. The problem is to learn the optimal actions so as to maximize the total reward. The learning problems differ in the details of how the data is collected and how performance is measured.

### 2.3.3.1 Markov Decision Processes

The environment is assumed to be stochastic. Further, the measurements available on the environment's state are detailed enough so that the agent can avoid reasoning about how to collect information about the state. Problems with these characteristics are best described in the framework of Markov Decision Processes (MDPs).

A MDP can defined as a triplet $\mathcal{M} = (\mathcal{S}, \mathcal{A}, \mathcal{P}_0)$, where $\mathcal{S}$ is the set of states, $\mathcal{A}$ is the set of actions, and $\mathcal{P}_0$ is the transition probability. $\mathcal{P}_0$ gives the probability of moving from state $s$ to some other state $s'$ provided that action $a$ was chosen in state $s$, which can be denoted by $\mathcal{P}(s, a, s')$.

In addition, $\mathcal{P}_0$ also gives rise to the immediate reward function $r : \mathcal{S} \times \mathcal{A} \rightarrow \mathbb{R}$, which gives the expected immediate reward received when action $a$ is chosen in state $s$: If $(S'_{(s,a)}; R_{(s,a)}) \sim \mathcal{P}_0(.|s, a)$, then $r(s, a) = \mathbb{E}[R_{(s,a)}]$.

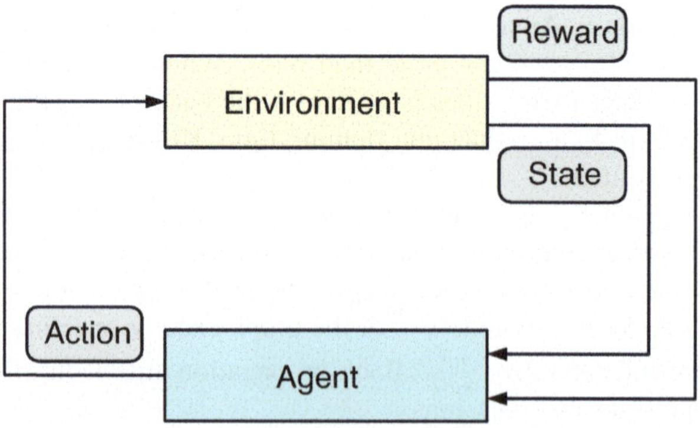


**Fig. 2.3** A basic illustration on reinforcement learning

MDPs are a tool for modeling sequential decision-making problems where a decision maker interacts with the environment in a sequential fashion. Given an MDP $\mathcal{M}$, this interaction happens as follows: Let $t$ denote the current time (or stage), let $S_t \in \mathcal{S}$ and $A_t \in \mathcal{A}$ denote the random state of the environment and the action chosen by the decision maker at time $t$, respectively. Once the action is selected, it is sent to the environment, which makes a transition:

$$(S_{t+1}, R_{t+1}) \sim \mathcal{P}_0(.|S_t, A_t). \tag{2.46}$$

In particular, $S_{t+1}$ is random and $P(S_{t+1} = s'|S_t = s, A_t = a) = \mathcal{P}(s, a, s')$ holds for any $s, y \in \mathcal{S}, a \in \mathcal{A}$. Further, $\mathbb{E}[R_{t+1}|S_t, A_t] = r(S_t, A_t)$. The agent then observes the next state $s_{t+1}$ and reward $R_{t+1}$, chooses a new action $A_{t+1}$ and the process is repeated. The goal of the agent is to choose some actions so as to maximize the expected total discounted reward.

The agent can select its actions at any stage based on the observed history. A rule describing the way to select actions is called a behavior. A behavior of the agent and some initial random state $S_0$ together define a random state-action-reward sequence $(S_t, A_t, R_{t+1})$, where $(S_{t+1}, R_{t+1})$ is connected to $(S_t, A_t)$ by (2.46), and $A_t$ is the action prescribed by the behavior based on the history. Thus, the return underlying a behavior is defined as

$$\mathcal{R} = \sum_{t=0}^{\infty} \gamma^t R_{t+1}, \tag{2.47}$$

where $0 \leq \gamma \leq 1$ is a discount factor. Here, $\gamma < 1$ implies that the rewards far in the future worth exponentially less than the reward received at the first stage. An MDP when the return is defined with $\gamma < 1$ is called discounted. The goal of the agent is to choose a behavior that maximizes the expected return, irrespectively of how the process is started. Such a maximizing behavior is said to be optimal.

The most straightforward way to find an optimal behavior in the MDP is to list all behaviors and then identify the ones that give the highest possible value for each initial state. However, there are too many behaviors, which makes this plan not viable. A better approach is based on value functions. In this approach, one first computes the so-called optimal value function, which then allows one to determine an optimal behavior with relative easiness. The optimal value, $V^*(s)$, for state $s \in \mathcal{S}$ gives the highest achievable expected return when the process is started from state $s$. The function $V^*$ is called the optimal value function. A behavior that achieves the optimal values in all states is optimal.

*Deterministic stationary policies* represent a special class of behaviors, which play an important role in the theory of MDPs. They are specified by some mapping $\pi$, which maps states to actions, i.e., $\pi : \mathcal{S} \rightarrow \mathcal{A}$. At any time $t \geq 0$, the action $A_t$ is selected by

$$A_t = \pi(S_t). \tag{2.48}$$

More generally, a *stochastic stationary policy* $\pi$ maps states to distributions over the action space. In such a policy $\pi$, we use $\pi(a|s)$ to denote the probability of action a being selected by $\pi$ in state $s$. Note that if a stationary policy is followed in an MDP, i.e.,

$$A_t \sim \pi(.|S_t), \tag{2.49}$$

the state process $(S_t, t \geq 0)$ will be a Markov chain. We will use $\Pi_{stat}$ to denote the set of all stationary policies. For brevity, we will often say "policy" instead of "stationary policy."

For policy $\pi \in \Pi_{stat}$, the value function, $V^\pi : \mathscr{S} \to \mathbb{R}$, underlying $\pi$ is defined by

$$V^\pi(s) = \mathbb{E}\left[\sum_{t=0}^{\infty} \gamma^t R_{t+1} | S_0 = s\right], s \in \mathscr{S}. \tag{2.50}$$

It will also be useful to define the action-value function, $Q^\pi : \mathscr{S} \times \mathscr{A} \to \mathbb{R}$, underlying a policy $\pi$ in an MDP: Let $(S_t, A_t, R_{t+1})$ be the resulting stochastic process, we have

$$Q^\pi(s, a) = \mathbb{E}\left[\sum_{t=0}^{\infty} \gamma^t R_{t+1} | S_0 = s, A_0 = a\right], s \in \mathscr{S}, a \in \mathscr{A}. \tag{2.51}$$

Similarly to $V^*(s)$, the optimal action-value $Q^*(s, a)$ at the state-action pair $(s, a)$ is defined as the maximum of the expected return under the constraints that the process starts at state $s$, and the first action selected is $a$. The optimal value- and action-value functions are connected by the following equations:

$$\begin{aligned} V^*(s) &= \sup_{a \in \mathscr{A}} Q^*(s, a), s \in \mathscr{S}; \\ Q^*(s, a) &= r(s, a) + \gamma \sum_{s' \in \mathscr{S}} P(s, a, s') V^*(s'), s \in \mathscr{S}, a \in \mathscr{A}. \end{aligned} \tag{2.52}$$

In the class of MDPs considered here, an optimal stationary policy always exists:

$$V^*(s) = \sup_{\pi \in \Pi_{stat}} V^\pi(s), s \in \mathscr{S}. \tag{2.53}$$

In fact, any policy $\pi \in \Pi_{stat}$ which satisfies the equality

$$\sum_{a \in \mathscr{A}} \pi(a|s) Q^*(s, a) = V^*(s) \tag{2.54}$$

simultaneously for all states $s \in \mathscr{S}$ is optimal.

The next question is how to find $V^*$ or $Q^*$. We have the following theorem on how to find the value function of a policy.

**Theorem 2.1 (Bellman Equations for Deterministic Policies)** *For an MDP* $\mathcal{M} = (\mathcal{S}, \mathcal{A}, \mathcal{P}_0)$, *a discount factor* $\gamma$ *and deterministic policy* $\pi \in \Pi_{stat}$. *Let* $r$ *be the immediate reward function of* $\mathcal{M}$. *Then* $V^\pi$ *satisfies*

$$V^\pi(s) = r(s, \pi(s)) + \gamma \sum_{s' \in \mathcal{S}} P(s, \pi(s), s')V^\pi(s'), s \in \mathcal{S} : \tag{2.55}$$

*This equation is called the Bellman equation for* $V^\pi$.

#### 2.3.3.2 Reinforcement Learning Methods

We now turn to how to learn the optimal policy. The goal of reinforcement learning is to find the optimal policies for the agents. At time $t$, the environment is at a certain state, and each agent selects a certain action according to its policy. Then, the environment then transits into a new state which is determined by its previous state and the actions of the agents. A reward is generated for each agent, which quantifies how well the objective of the agent is achieved.

Typically, the reinforcement learning methods can broadly categorized into four types: multi-armed bandit learning, Q-learning, actor-critic learning, and deep reinforcement learning. In the following, we will elaborate on these four methods [14].

1. **Multi-armed Bandit Learning**: As shown in Fig. 2.4a, we consider an MDP that has a single state. Let the problem be that of maximizing the return while learning. Since there is only one state, this is an instance of the classical bandit problems. A basic observation is that the agent who always chooses the action with the best estimated payoff (i.e., who always makes the *greedy* choice) can fail to find the best action with positive probability, which in turn leads to a large loss. Thus, the agent must take actions to explore. The question is then how to balance the exploration and the exploitation.

   A simple strategy is to fix $\epsilon > 0$ and choose a randomly selected action with probability $\epsilon$, and go with the optimal choice otherwise. Another simple strategy is the so-called Boltzmann exploration strategy. Given $Q_t(a)$ of action $a \in \mathcal{A}$ at time $t$, the next action is drawn from the following distribution:

$$\pi(a) = \frac{\exp(\beta Q_t(a))}{\sum_{a' \in \mathcal{A}} \exp(\beta Q_t(a'))}. \tag{2.56}$$

   Here, $\beta > 0$ controls the greediness of action selection ($\beta \to \infty$ results in a greedy choice). The difference between Boltzmann exploration and $\epsilon$-greedy is that $\epsilon$-greedy does not take into account the relative values of the actions, while Boltzmann exploration does.

   Since the different states of environment have not been considered, the multi-armed bandit learning is inefficient to deal with environments with

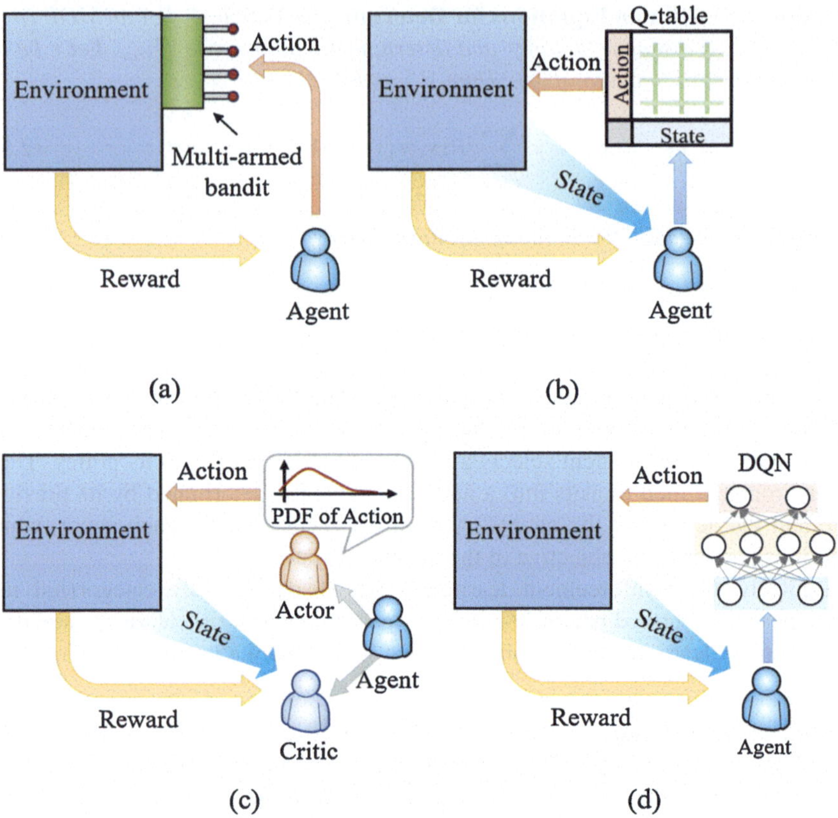


**Fig. 2.4** Illustration of (**a**) multi-armed bandit learning (**b**) Q-learning, (**c**) actor-critic learning, and (**d**) deep reinforcement learning

rapid state changes. However, it is compensated by its very low complexity in implementation.

2. **Q-Learning**: Q-learning algorithms are to approximate the optimal action-value function $Q^*$ directly. The Q-learning algorithms can be thought of as approximate versions of value iteration that generate some sequence of action-value functions $Q_k$. The idea is that if $Q_k$ is close to $Q^*$, the policy that is greedy with respect to $Q_k$ will be close to optimal.

   Fix a finite MDP $\mathscr{M} = (\mathscr{S}, \mathscr{A}, \mathscr{P}_0)$ and a discount factor $\gamma$. The Q-learning algorithm keeps an estimate $Q_t(s, a)$ of $Q^*(s, a)$ for each state-action pair $(s, a) \in \mathscr{S} \times \mathscr{A}$, which is stored in a Q-table, as shown in Fig. 2.4b. Upon observing $(S_t, A_t, R_{t+1}, S'_{t+1})$, the estimates are updated as follows:

$$\begin{aligned} \delta_{t+1}(Q) &= R_{t+1} + \gamma \max_{a' \in \mathscr{A}} Q(S'_{t+1}, a') - Q(S_t, A_t), \\ Q_{t+1}(s, a) &= Q_t(s, a) + \alpha_t \delta_{t+1}(Q_t) \mathbb{I}_{s=S_t, a=A_t}, \ (s, a) \in \mathscr{S} \times \mathscr{A}. \end{aligned} \tag{2.57}$$

Here, $(S'_{t+1}, R_{t+1}) \sim \mathscr{P}_0(.|S_t, A_t)$. In stochastic equilibrium, one must have $\mathbb{E}[\delta_{t+1}(Q)|S_t = s, A_t = a] = 0$ for any $(s, a) \in \mathscr{S} \times \mathscr{A}$.

3. **Actor-Critic Learning**: Actor-critic methods implement generalized policy iterations. When using sample-based methods or function approximation, exact evaluation of the policies may require infinitely many samples or might be impossible due to the restrictions of the function-approximation technique. Hence, reinforcement learning algorithms simulating policy iteration must change the policy based on incomplete knowledge of the value function.

   Algorithms that update the policy before it is completely evaluated are said to implement generalized policy iteration (GPI). In GPI, there are two closely interacting processes of an actor and a critic: the actor aims at improving the current policy, while the critic evaluates the current policy, thus helping the actor. The interaction of the actor and the critic is illustrated on Fig. 2.4c.

   In the following, we will describe value estimation methods (used by the critic) and policy improvement methods (used by the actor), respectively.

*Critic* The job of the critic is to estimate the value of the current target policy of the actor. The critic performs a temporal-difference (TD) algorithm with linear parameterized approximation for the Q-function, which can be given by

$$
\begin{aligned}
\delta_{t+1} &= R_{t+1} + \gamma V_{\theta_t}(S_{t+1}) - V_{\theta_t}(S_t), \\
z_{t+1} &= \nabla_\theta V_{\theta_t}(S_t) + \gamma \lambda z_t, \\
\theta_{t+1} &= \theta_t + \alpha_t \delta_{t+1} z_{t+1} \\
z_0 &= 0,
\end{aligned}
\tag{2.58}
$$

where $\theta \in \mathbb{R}^d$ is the parameter of the function approximation and $z \in \mathbb{R}^d$ is the vector of eligibility traces.

Here, if the policy $\pi$ is a stochastic policy, $V_{t+1}$ can be given by

$$
V_{t+1} = \sum_{a \in \mathscr{A}} \pi(a|S'_{t+1}) Q_\theta(S'_{t+1}, a). \tag{2.59}
$$

If $\pi$ is a deterministic policy, this equation can be simplified to $V_{t+1} = Q_\theta(S'_{t+1}, \pi(S'_{t+1}))$.

When the action space is large or continuous and stochastic policies are considered, evaluating the sums $\sum_{a \in \mathscr{A}} \pi(a|s)\psi(s, a)$ might be infeasible. One solution is to introduce some state features $\phi : \mathscr{S} \to \mathbb{R}^d$ so that $\sum_{a \in \mathscr{A}} \pi(a|s)\phi(s, a) = 0$ holds for any state $s \in \mathscr{S}$. Then, define $Q_\theta(s, a) = \theta^T(\psi(s, a) + \phi(s, a))$, and $V_\theta(s) = \sum_{a \in \mathscr{A}} \pi(a|s) Q_\theta(s, a) = \theta^T \psi(s)$.

*Actor* The idea of the policy improvement is to perform gradient ascent directly on the performance surface underlying a chosen parametric policy class. These methods perform stochastic gradient ascent on the performance surface induced by a smoothly parameterized policy class $\Pi = (\pi_\omega)$ of stochastic stationary policies.

Given $\Pi$, the problem is to find the value of $\omega$ corresponding to the best policy

$$\arg\max_{\omega} \rho_\omega. \tag{2.60}$$

Here, $\rho_\omega$ can be measured by the expected return of policy $\pi_\omega$, with respect to some initial distribution over the states. The initial distribution can be the stationary distribution underlying the policy chosen, in which case maximizing $\rho_\omega$ will be equivalent to maximizing the long-run average reward.

Assume that the Markov chain resulting from policy $\pi_\omega$ is ergodic, regardless of the value of $\omega$. The question is how to estimate the gradient of $\pi_\omega$. Let $\phi_\omega(s, a) : \mathscr{S} \times \mathscr{A} \rightarrow \mathbb{R}^{d_\omega}$ be the function underlying $\pi_\omega$:

$$\phi_\omega(s, a) = \frac{\partial}{\partial \omega} \log \pi_\omega(a|s). \tag{2.61}$$

Also, define

$$G(\omega) = (Q^{\pi_\omega}(S, A) - h(S))\phi_\omega(S, A). \tag{2.62}$$

Here, $(S, A)$ is a sample from the stationary state-action distribution underlying policy $\pi_\omega$. $Q^{\pi_\omega}$ is the action-value function of $\pi_\omega$ and $h(.)$ is an arbitrary bounded function. Therefore, $G(\omega)$ is an unbiased estimate of the gradient:

$$\nabla_\omega \rho_\omega = \mathbb{E}[G(\omega)]. \tag{2.63}$$

Let $(S_t, A_t)$ be a sample from the stationary distribution underlying $\pi_{\omega_t}$. Therefore, $G(\omega)$ can be obtained by the following rule:

$$\begin{aligned} \hat{G}_t &= (\hat{Q}_t(S_t, A_t) - h(S_t))\phi_{\omega_t}(S_t, A_t), \\ \omega_{t+1} &= \omega_t + \beta_t \hat{G}^t. \end{aligned} \tag{2.64}$$

Here, $\beta_t$ is a constant, and

$$\mathbb{E}[\hat{Q}_t(S_t, A_t)\phi_{\omega_t}(S_t, A_t)] = \mathbb{E}[\hat{Q}^{\pi_{\omega_t}}(S_t, A_t)\phi_{\omega_t}(S_t, A_t)]. \tag{2.65}$$

The role of $h(.)$ function is to reduce the variance of the gradient estimate $\hat{G}^t$ so as to speed up the convergence.

4. **Deep Reinforcement Learning**: Deep reinforcement learning is proposed to scale up the prior work in reinforcement learning to the high-dimensional problems, due to the powerful function approximation properties of neural networks as we have introduced in the last subsection. Therefore, deep reinforcement learning based on training deep neural networks can effectively approximate the optimal policy $\pi^*$ or the optimal value functions $V^*$ or $Q^*$.

   The most widely used deep reinforcement learning approach is the deep Q-network (DQN), where we utilize a neural network to approximate the value of $Q^*$ [12], as shown in Fig. 2.4d. The DQN can address the fundamental instability problem of using function approximation in RL by the use of two techniques: experience replay and target networks. Experience replay memory stores transitions of the form $(\boldsymbol{s}_t, \boldsymbol{a}_t, \boldsymbol{s}_{t+1}, \boldsymbol{r}_{t+1})$ in a cyclic buffer, which enables the agent in the reinforcement learning to sample from and train on previously observed data offline. This can not only massively reduce the number of interactions with the environment, but also the batches of experience can be sampled, and thus reducing the variance of learning updates. From a practical perspective, batches of data can be efficiently processed in parallel by modern hardware, increasing throughput.

   One of the key components of the DQN is a function approximator for the Q-function $Q(s, a; \theta)$ with parameter $\theta$. To estimate this network, we optimize the following sequence of loss functions at iteration $i$ [13]:

$$L_i(\theta_i) = \mathbb{E}_{s,a,r,s'}\left[(y_i^{DQN} - Q(s, a; \theta_i))^2\right] \tag{2.66}$$

   with

$$y_i^{DQN} = r + \gamma \max_{a' \in \mathscr{A}} Q(s', a'; \theta^-)), \tag{2.67}$$

   where $\theta^-$ represents the parameters of a fixed and separate target network. In practice, an important trick is to freeze the parameters of the target network $Q(s', a'; \theta^-)$ for a fixed number of iterations while updating the online network $Q(s, a; \theta_i)$ by gradient descent, which can greatly improves the stability of the algorithm. The specific gradient update can be written by

$$\nabla_{\theta_i} L_i(\theta_i) = \mathbb{E}_{s,a,r,s'}\left[(y_i^{DQN} - Q(s, a; \theta_i))\nabla_{\theta_i} Q(s, a; \theta_i)\right]. \tag{2.68}$$

   This approach is model free in the sense that the states and rewards are produced by the environment.

In summary, the characteristics of these four reinforcement learning methods are given in Table 2.2.

**Table 2.2** Summary of reinforcement learning methods

| Reinforcement learning methods | Characteristics |
|---|---|
| Multi-armed bandit learning | Very low complexity |
| | Consider one state |
| Q-learning | Low complexity |
| | Consider state transitions |
| | Inefficient to deal with large action spaces |
| Actor-critic learning | Low complexity |
| | Consider state transitions |
| | Capability to handle large or continuous action spaces |
| Deep reinforcement learning | High complexity |
| | Consider state transitions |
| | Capability to handle large state spaces |

# References


1. S. Bubeck, Convex optimization: algorithms and complexity. Found. Trends Mach. Learning **8**(3–4) (2015)
2. S. Boyd, L. Vandenberghe, *Convex Optimization* (Cambridge University Press, Cambridge, 2004)
3. L.T.H. An, P.D. Tao, The DC (difference of convex functions) programming and DCA revisited with DC models of real world nonconvex optimization problem. Ann. Oper. Res. **133**(1–4), 23–46 (2005)
4. D. Li, X. Sun, *Nonlinear Integer Programming* (Springer Science & Business Media, New York, 2006)
5. Z. Han, D. Niyato, W. Saad, T. Başar, A. Hjørungnes, *Game Theory in Wireless and Communication Networks: Theory, Models, and Applications* (Cambridge University Press, Cambridge, 2012)
6. P. Bolton, M. Dewatripont, *Contract Theory* (The MIT Press, Cambridge, 2004)
7. Y. Zhang, Z. Han, *Contract Theory for Wireless Networks* (Springer, Switzerland, 2017)
8. T.M. Mitchell, *Machine Learning* (McGraw-Hill, New York, 1997)
9. I. Goodfellow, Y. Bengio, A. Courville, *Deep Learning* (The MIT Press, Cambridge, 2016)
10. D. Rumelhart, G. Hinton, R. Williams, Learning representations by back-propagating errors. Nature **323**, 533–536 (1986)
11. C. Szepesvári, Algorithms for reinforcement learning. Synth. Lect. Artif. Intell. Mach. Learn. **4**(1), 1–103 (2010)
12. K. Arulkumaran, M.P. Deisenroth, M. Brundage, A.A. Bharath, Deep reinforcement learning: a brief survey. IEEE Signal Process. Mag. **34**(6), 26–38 (2017)
13. Z. Wang, T. Schaul, M. Hessel, H.V. Hasselt, M. Lanctot, N. de Freitas, Dueling network architectures for deep reinforcement learning, in *Proceedings of ACM ICMLR*, New York (2016)
14. J. Hu, H. Zhang, L. Song, Z. Han, H.V. Poor, Reinforcement learning for a cellular internet of UAVs: protocol design, trajectory control, and resource management. IEEE Wireless Commun. Early Access. https://arxiv.org/abs/1911.08771

# Chapter 3
# UAV Assisted Cellular Communications

Dedicated UAVs can be used as communication platforms in the way as wireless access points or relays nodes, to further assist the terrestrial communications. This type of applications can be referred to as **UAV assisted cellular communications**. UAV assisted cellular communications have numerous use cases, including traffic offloading, wireless backhauling, swift service recovery after natural disasters, emergency response, rescue and search, information dissemination/broadcasting, and data collection from ground sensors for machine-type communications. However, different from traditional cellular networks, how to plan the time-variant placements of the UAVs served as BS/relay is very challenging due to the complicated 3D propagation environments as well as many other practical constraints such as power and flying speed. In addition, spectrum sharing with existing cellular networks is another interesting topic to investigate.

In this chapter, we first discuss the UAV offloading problem where the UAVs serve as BSs in Sect. 3.1, and then optimize the trajectory and transmit power in the UAV relay networks in Sect. 3.2.

## 3.1 UAVs Serving as Base Stations

The rapid development of wireless communication enabled small-scale UAVs has created a variate of civil applications [1], from cargo delivery [2] and remote sensing [3] to data relaying [4] and connectivity maintenance [5, 6]. From the aspect of wireless communications, one major advantage of utilizing UAVs is their high probability of keeping LoS signals with other communication nodes, alleviating the problem brought by severe shadowing in urban or mountainous terrain [7, 8]. Different from high-altitude platforms which are designed for long-term assignment above tens of kilometers height [9], small-scale UAVs within only hundreds of meters off the ground can be deployed more quickly. In addition, the properties



H. Zhang et al., *Unmanned Aerial Vehicle Applications over Cellular Networks for 5G and Beyond*, Wireless Networks, https://doi.org/10.1007/978-3-030-33039-2_3

like low cost, high flexibility, and ease of scheduling also make small-scale UAVs a favorable choice in civil usages, in spite of their disadvantages such as low battery capacity [10].

One of the major problems in the UAV assisted wireless communications is to optimally deploy UAVs, in which way mobile users can be better served [10]. Many studies have been done to deal with this problem from distinctive viewpoints with respect to different objectives and constraints [11–23]. Among them, the works in [11–14] considered the scenario consisting only one UAV to provide with coverage, the works in [15–18] took into account multiple UAVs to providing better services by joint coverage, and the works in [19–23] studied the coexistence of BSs and multi-UAVs, where data offloading becomes a major problem.

To be specific, in [11], the optimal height of a single UAV was deduced to maximize the coverage radius. The authors of [12] minimized the transmission power of the UAV with fixed coverage radius. The problem of maximizing the number of users covered by one UAV is studied in [13]. And the authors of [14] further took into account the interference from D2D users. For multiple UAVs, the coverage probability of a ground user was derived in [15]. The work in [16] proposed a solution to minimize the number of UAVs to cover all the users. The authors of [17] studied the deployment of multiple UAVs to achieve largest total coverage area. And in [18], the total transmission power of UAVs was minimized while the data rate for each user was guaranteed. With the consideration of BSs in the scenario, the gain of deploying additional UAVs for offloading was discussed in [19–21]. The authors of [22] focused on the optimal cell partition strategy to minimize average delay of the users in a cellular network with multiple UAVs. In [23], the optimal resource allocation was presented, where one macro-cell base station (MBS), multiple small-cell base stations (SBSs), and multiple UAVs are involved.

Although UAV coverage and offloading problems have been widely discussed, few existing studies consider the situation where UAV operators could be selfish individuals with different objectives [24]. For instance, the venue owners and scenic area managers may want to temporarily deploy their own UAVs to better serve their visitors, due to the temporarily increased number of mobile users or the inconvenience of installing SBSs in remote areas [25]. In such cases, the deployment of multiple UAVs depends on each UAV operator, and the solution is not likely to be optimal as calculated by centralized algorithms. In addition, the wireless channel allocation becomes a more critical problem since the bandwidth that the UAVs used to serve mobile users has to be explicitly authorized by the MBS manager. Therefore, further studies need to be done with respect to selfish UAV operators in UAV assisted offloading cellular networks.

In this section, we focus on the scenario with one MBS managed by the MBS manager, and multiple SBS-enabled UAVs owned by different UAV operators. To enable downlink transmissions of the UAVs, each UAV operator has to buy a certain amount of bandwidth authorized by the MBS manager. However, the total usable bandwidth of the MBS is limited, and selling part of the total bandwidth to the UAVs may harm the capacity of the MBS. Therefore, payments to the MBS manager should be made by UAV operators. Here, contract theory [26] can be applied as a tool to analyze the optimal contract that the MBS manager will design to maximize its revenue. Specifically, such contract comprises a set of bandwidth options and corresponding prices. Since each UAV operator only chooses the most profitable option from the whole contract, the MBS manager has to guarantee that the contract is feasible, i.e., the option that a UAV operator chooses from the contract is exactly the one that designed for it.

The rest of this section is organized as follows. Section 3.1.1 presents our system model and formulates the optimal contract design problem. Section 3.1.2 theoretically deduces the optimal solution and provides our dynamic programming algorithm. Section 3.1.3 focuses on the height of the UAVs and discusses its impact on the revenue of the MBS manager. Section 3.1.4 shows the simulation results of the optimal contract. Finally, we summarize this section in Sect. 3.1.5.

### *3.1.1 System Model*

We consider a scenario with one MBS and $N$ UAVs, as shown in Fig. 3.1 [27]. The MBS is operated by a MBS manager and the UAVs are run by different UAV operators. All the UAVs has to stay at a legal height $H$, which is designated by the MBS manager. While the horizontal location of each UAV can be adjusted by its operator to cover as many local users as possible. Each UAV operator aims to provide better services for its local mobile users with licensed spectrum, which is temporally bought from the MBS manager.

In the rest part of this subsection, we first discuss the concerned mobility and energy consumption of the UAVs, then present the wireless downlink model of the MBS and the UAVs. After that, we introduce the utility of the UAV operators as well as the cost of the MBS manager, and finally formulate the contract design problem.

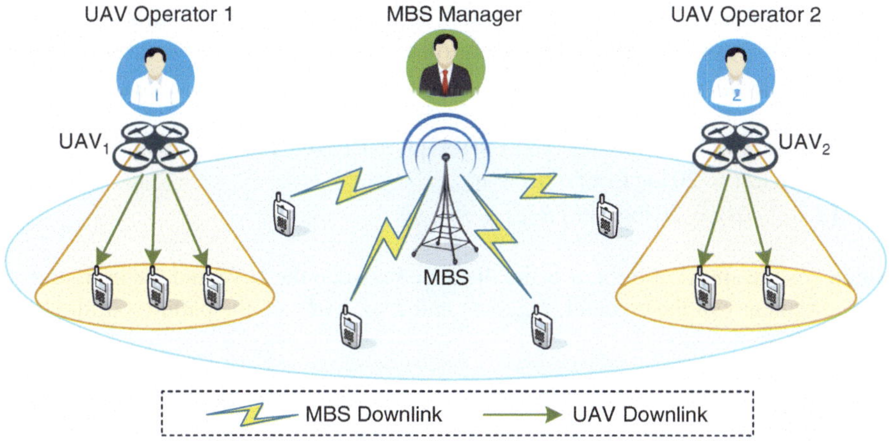


**Fig. 3.1** The system model of UAV assisted offloading in cellular networks with one MBS manager and multiple UAV operators

#### 3.1.1.1 Mobility and Energy Consumption

Without the loss of generality, we consider the UAV offloading system in a series of short time slots.[1] In the $s$th time slot, the distribution of mobile users as well as the horizontal location of each UAV is assumed to be stable. In the $(s+1)$th time slot, the horizontal location of a UAV can be adjusted by its operator, to cover as many of its own mobile users as possible.

The total available energy for the $n$th UAV to stabilize or adjust its location is denoted by $E_n$. The energy consumption of UAV $n$ keeping itself stabilized for a whole time slot is given by $e_n$. The additional energy consumption of moving UAV $n$ for a distance of $l$ between time slots is denoted by $q_n \cdot l$, where $q_n$ is a constant for UAV $n$. With $E_n$, $e_n$, $q_n$, and a specific movement behavior, we are able to obtain the number of time slots that UAV $n$ could sustain to provide wireless connections for its users.[2]

For the $n$th UAV operator, we also assume that there is a constant cost of deploying and retrieving UAV $n$ (unrelated to the number of time slots), denoted by $C_n$. To make it worth deploying its UAV, this UAV operator has to maximize his profit in each time slot during the deployment. In addition, the MBS manager also aims to maximize its own revenue in each time slot by properly designing the contract. Since the following parts of this subsection only correspond to the problem within one time slot, we omit the time slot number $s$ for reading convenience.

#### 3.1.1.2 Wireless Downlink Model

The air-to-ground wireless channel between a UAV and a mobile user mainly consists of two parts, which are the LoS component and the NLoS component [8]. Based on the study in [11], the probability of LoS for a user with elevation angle $\theta$ (in degree) to a specific UAV is given by $P_{LoS}(\theta) = \frac{1}{1+a\exp\left(-b[\theta-a]\right)}$, where $a$ and $b$ are the parameters that depend on the specific terrain (like urban, rural, etc.).

Based on $P_{LoS}$, the average pathloss from the UAV to the user can be given by (in dB)

$$\begin{cases} \overline{L}_{UAV}(\theta,d) = P_{LoS}(\theta)\cdot L_{LoS}(d) + \left[1 - P_{LoS}(\theta)\right]\cdot L_{NLoS}(d), \\ L_{LoS}(d) = 20\log\left(4\pi f d/c\right) + \eta_{LoS}, \\ L_{NLoS}(d) = 20\log\left(4\pi f d/c\right) + \eta_{NLoS}, \end{cases} \tag{3.1}$$

where $c$ is the speed of light, $d$ is the distance between the UAV and the user, and $f$ is the frequency of the channel. $L_{LoS}(d)$ and $L_{NLoS}(d)$ are the pathloss of the LoS

[1] The length of each time slots can be around a few seconds since our algorithm has a high efficiency.

[2] The power consumption of wireless transmission can be ignored compared to that of the engines of the UAV [4].

component and the pathloss of the NLoS component, respectively. $\eta_{LoS}$, $\eta_{NLoS}$ are the average additional loss that depends on the environment. In contrast to the UAV-to-user wireless channel, the MBS-to-user channels are considered as NLoS only, which gives us the average pathloss as[3]

$$\overline{L}_{MBS}(d) = 20 \log (4\pi f d/c) + \eta_{NLoS}. \tag{3.2}$$

For simplicity, we assume that different channels has a similar $f$ and the difference can be ignored.

To see the signal quality that each user could experience, we use $\gamma_{MBS}(d)$ to denote the signal-to-noise ratio (SNR) for MSB users at the distance $d$ from the MBS. And we have $\gamma_{MBS}(d) = \left[P_{MBS} - \overline{L}_{MBS}(d)\right]/N_0$, where $P_{MBS}$ is the transmission power of the MBS and $N_0$ is the power of background noise. Similarly, we use $\gamma_{UAV}(d, \theta)$ to denote the SNR for the UAV users with elevation angle $\theta$ and distance $d$ from a certain UAV, given as $\gamma_{UAV}(d, \theta) = \left[P_{UAV} - \overline{L}_{UAV}(d, \theta)\right]/N_0$, where $P_{UAV}$ is the transmission power of the UAV.

It is also assumed that each user can automatically choose among the MBS and the UAVs to obtain the best SNR. Therefore, it is necessary to find out in which region a certain UAV is able to provide better SNR than the others (including the MBS and the other UAVs). We denote the region where UAV$_n$ provides better SNR as UAV$_n$'s effective offloading region, denoted by $\Omega_n$.

#### 3.1.1.3 The Utility of the UAV Operators

Each mobile user in an effective offloading region is assumed to access the UAV randomly. We call the number of the users in $\Omega_n$ that want to connect to UAV $n$ at any instant as the "active user number" of UAV $n$, denoted by $\varepsilon_n$. We assume that $\varepsilon_n$ obeys Poisson distribution[4] with mean value of $\mu_n$. Based on $\mu_n$, we can classify the UAVs into multiple types. Specifically, we refer to UAV $n$ as a $\lambda$-type UAV if $\mu_n = \lambda$, which means that there are averagely $\lambda$ users connecting to UAV$_n$ at any instant. The number of $\lambda$-type UAVs is denoted by $N_\lambda$, where $\sum_\lambda N_\lambda = N$. For writing simplicity, we use random variable $X_\lambda$ (instead of $\varepsilon_n$) to denote the active user number of a $\lambda$-type UAV. The probability of $X_\lambda = k$ is given by

$$P(X_\lambda = k) = \frac{(\lambda)^k}{k!} e^{-\lambda}, \qquad k = 0, 1, 2, \cdots \tag{3.3}$$

Without the loss of generality, we assume that each mobile user connecting to a UAV (or the MBS) is allocated with one channel with fixed bandwidth $B$, in a

[3]Small-scale fading is ignored, since we only use average SNR to determine which UAV or MBS the user should connect to.

[4]We assume that each potential user has an independent probability to become an active user at each moment. Therefore, the number of active users in a certain area should be a binomial distributed random variable. When the number of users in each area is large enough (>100), Poisson distribution is a good approximation of binomial distribution without the loss of accuracy.

frequency division pattern. Due to the variation of the active user number, there is always a probability that an UAV fails to serve the current active users. Therefore, the more channels are obtained, the more utility the UAV can achieve. The utility function of obtaining $w$ channels for a $\lambda$-type UAV is denoted by $U(\lambda, w)$. Since the utility of obtaining no channels is 0, we have

$$U(\lambda, w) = 0, \qquad w = 0. \tag{3.4}$$

Now assume that we have determined the value of $U(\lambda, w-1)$, the rest problem is how to obtain $U(\lambda, w)$ by figuring out the marginal utility of obtaining the $w$th channel. Note that the newly added $w$th channel is only useful when there are more than $w-1$ active users at the given moment. Therefore, the marginal utility is $P(X_\lambda \geq w) \times 1$, i.e., the probability of more than $w-1$ users is active at the moment. Thus we have

$$U(\lambda, w) = U(\lambda, (w-1)) + P(X_\lambda \geq w), \qquad w \geq 1. \tag{3.5}$$

Based on (3.4) and (3.5), we can derive the general term of the $\lambda$-type UAV's utility as

$$U(\lambda, w) = \sum_{k=1}^{k=w} P(X_\lambda \geq k), \qquad w \geq 1. \tag{3.6}$$

#### 3.1.1.4 Cost of the MBS Manager

It is assumed that the MBS will not reuse the spectrum that has already been sold, which implies the MBS manager suffers a certain degree of loss as it sells the spectrum to UAV operators. The active user number of the MBS is also assumed to follow the Poisson distribution. We denote this random variable as $X_{BS}$ and the mean value of it as $\lambda_{BS}$. Therefore, we have

$$P(X_{BS} = k) = \frac{(\lambda_{BS})^k}{k!} e^{-\lambda_{BS}}, \qquad k = 0, 1, 2, \cdots \tag{3.7}$$

The total number of channels of the MBS is denoted by $M$, $M \in \mathbb{Z}^+$. Just like the situation of UAVs, there is also a utility of a certain number of channels for the MBS manager, $U_{BS}(m)$, representing the average number of users that $m$ channels can serve, given as $U_{BS}(m) = 0$, for $m = 0$, and $U_{BS}(m) = U_{BS}(m-1) + P(X_{BS} \geq m)$, for $m \geq 1$. Based on the utility of the MBS manager, we define the cost function $C(m)$ as the utility loss of reducing the number of channels from $M$ to $M-m$, given as

$$C(m) = U_{BS}(M) - U_{BS}(M-m) = \sum_{k=M-m+1}^{M} P(X_{BS} \geq k). \tag{3.8}$$

#### 3.1.1.5 Contract Formulation

Since different types of UAVs have different demands, the MBS manager has to design a contract which contains a set of "quality-price" options for all the UAV operators, denoted by $\left\{(w(\lambda), p(\lambda)) \mid \forall \lambda \in \Lambda\right\}$, where $\Lambda$ represents the set of all the UAV types. In this contract, the quality $w(\lambda)$ is the number of channels designed to sell to a $\lambda$-type UAV operator, and $p(\lambda)$ is the corresponding price designed to be charged. Each $(w(\lambda), p(\lambda))$ pair can be seen as a commodity with quality $w(\lambda)$ at price $p(\lambda)$.

However, each UAV operator is expected to choose the one that maximizes its own profit according to the whole contract. The contract is feasible if and only if any $\lambda$-type UAV operator considers the commodity $(w(\lambda), p(\lambda))$ as its best choice. And to achieve this, the first requirement is the IC condition, implying that the commodity designed for a $\lambda$-type UAV operator in the contract is no worse than other commodities, given by

$$U\big(\lambda, w(\lambda)\big) - p(\lambda) \geq U\big(\lambda, w(\lambda')\big) - p(\lambda'), \qquad \forall \lambda' \neq \lambda. \tag{3.9}$$

If (3.9) is not satisfied, then a $\lambda$-type UAV operator may turn to another commodity, and the $\lambda$-type commodity is not properly designed. The second requirement is the IR condition, meaning that the $\lambda$-type UAV operator will not buy any of the commodities in the contract if all of the options lead to negative profits. In other words, the commodity designed for a $\lambda$-type UAV should lead to a nonnegative profit, even if this commodity is an "empty commodity" (with zero quality and zero price), given by

$$U\big(\lambda, w(\lambda)\big) - p(\lambda) \geq U\big(\lambda, 0\big) - 0 = 0, \tag{3.10}$$

where $U(\lambda, 0) - 0$ implies an "empty commodity" in the contract. This condition is added to avoid the case where the best commodity for a $\lambda$-type UAV is negative. In conclusion, a feasible contract has to satisfy the IC constraint and the IR constraint, and any contract that satisfies the IC and IR constraints is guaranteed to be feasible [28].

For the MBS manager, the overall revenue brought by the contract $\{w(\lambda), p(\lambda) \mid \forall \lambda \in \Lambda\}$ is

$$R = \sum_{\lambda \in \Lambda} \big(N_\lambda \cdot p(\lambda)\big) - C\Big(\sum_{\lambda \in \Lambda} N_\lambda \cdot w(\lambda)\Big), \tag{3.11}$$

where $N_\lambda \cdot p(\lambda)$ is the total payment obtained from $\lambda$-type UAV operators, and $\sum_{\lambda \in \Lambda} N_\lambda \cdot w(\lambda)$ is the total number of channels being sold. The objective of the MBS manager is to design proper $w(\lambda)$ and $p(\lambda)$ for any given $\lambda \in \Lambda$, in which way it can maximize its own revenue with the pre-consideration of each UAV operator's behavior, given as

**Table 3.1** Notations in our model

| | |
|---|---|
| $E_n$ | Total energy of UAV$_n$ |
| $e_n$ | Stabilization energy consumption of UAV$_n$ during a time slot |
| $q_n$ | Mobility energy consumption of UAV$_n$ for a unit distance |
| $C_n$ | Cost of deploying and retrieving UAV$_n$ |
| $a, b$ | Terrain parameters |
| $\eta_{NOLS}, \eta_{LOS}$ | Additional pathloss parameters for non-LoS and LoS |
| $P_{MBS}, P_{UAV}$ | Transmission power of the MBS and the UAVs |
| $\bar{L}_{UAV}(\theta, d), \bar{L}_{MBS}(d)$ | Average pathloss to the user with elevation angle $\theta$ and distance $d$ |
| $\Omega_n, S_n$ | Effective coverage region and effective coverage area of UAV$_n$ |
| $\varepsilon_n$ | Active user number of UAV$_n$ (random variable) |
| $\mu_n$ | Average active user number of UAV$_n$ |
| $\lambda$ | The type of a UAV (= its average active user number) |
| $\Lambda$ | The set of the types of the UAVs |
| $\lambda_{BS}$ | Average active user number of the MBS |
| $N_\lambda$ | Number of $\lambda$-type UAVs |
| $U(\lambda, w)$ | Utility of $w$ channels for a $\lambda$-type UAV |
| $M$ | Number of MBS's channels |
| $C(m)$ | Cost of the MBS when selling $m$ channels |
| $w(\lambda), p(\lambda)$ | Number of channels and corresponding price designed for a $\lambda$-type UAV |

$$
\begin{aligned}
\hat{R} = \max_{\{w(\lambda)\},\{p(\lambda)\}} & \sum_{\lambda\in\Lambda} \Big(N_\lambda \cdot p(\lambda)\Big) - C\Big(\sum_{\lambda\in\Lambda} N_\lambda \cdot w(\lambda)\Big), \\
s.t.\ \ & U\big(\lambda, w(\lambda)\big) - p(\lambda) \geq U\big(\lambda, w(\lambda')\big) - p(\lambda') \geq 0, && \forall \lambda, \lambda' \in \Lambda \text{ and } \lambda' \neq \lambda, \\
& U\big(\lambda, w(\lambda)\big) - p(\lambda) \geq 0, && \forall \lambda, \lambda' \in \Lambda \text{ and } \lambda' \neq \lambda, \\
& p(\lambda) \geq 0, \quad w(\lambda) = 0, 1, 2\cdots && \forall \lambda \in \Lambda, \\
& \sum_{\lambda\in\Lambda} N_\lambda \cdot w(\lambda) \leq M,
\end{aligned} \tag{3.12}
$$

where the first two constraints represent the IC and the IR, and the last one indicates the limited number of channels possessed by the MBS. In the rest part of this section, the quality assignment $w(\lambda)$, and the pricing strategy $p(\lambda)$, are the two most basic concerns. In addition, we call the contract that optimizes the problem in (3.12) as the "MBS optimal contract." Before studying the contact design problem, we provide Table 3.1 to summarize the notations in our model.

### 3.1.2 Optimal Contract Design

In this subsection, we exploit some basic properties of our problem in Sect. 3.1.2.1. By utilizing these properties, we provide the optimal pricing strategy based on the fixed quality assignment in Sect. 3.1.2.2. Next, we analyze and transform the optimal quality assignment problem in Sect. 3.1.2.3, in which way it can be solved by the proposed dynamic programming algorithm given in Sect. 3.1.2.4. And finally we discuss the socially optimal contract in Sect. 3.1.2.5.

To facilitate writing, we put all the types $\{\lambda\}$ in the ascending order, given by $\{\lambda_1, \cdots \lambda_t, \cdots \lambda_T\}$ where $T$ is the number of different types. We have $1 \leq t \leq T$ and $\lambda_{t_1} < \lambda_{t_2}$ if $t_1 < t_2$. Note that, in this case we call $\lambda_{t_1}$ as a "lower type" and $\lambda_{t_2}$ as a "higher type." In addition, we also simplify $N_{\lambda_t}$ as $N_t$, $w(\lambda_t)$ as $w_t$ and $p(\lambda_t)$ as $p_t$ in the discussions below.

#### 3.1.2.1 Basic Properties

Before we analyze the property of the utility function $U(\lambda, w)$, we first have to provide a more basic conclusion with respect to a property of Poisson distribution, on which the utility function is defined.

**Lemma 3.1** *Given that $X_\lambda$ and $X_{\lambda'}$ are two Poisson distribution random variables with mean values $\lambda$ and $\lambda'$, respectively, if $\lambda > \lambda' > 0$, then $P(X_\lambda \geq k) > P(X_{\lambda'} \geq k)$ for any $k \in \mathbb{Z}^+$.*

***Proof*** Consider $X_\alpha$ as a Poisson distribution random variable with mean value $\alpha$, we have $P(X_\alpha \geq k) = 1 - P(X_\alpha < k) = 1 - e^{-\alpha} \sum_{i=0}^{k-1} \frac{\alpha^i}{i!}$. Since $\alpha$ can be a real number in its definition domain, we derive the derivative of $P(X_\alpha \geq k)$ with respect to $\alpha$, given as

$$\frac{\partial P(X_\alpha \geq k)}{\partial \alpha} = e^{-\alpha} \sum_{i=0}^{k-1} \frac{\alpha^i}{i!} - e^{-\alpha} \frac{\partial}{\partial \alpha} \left( \sum_{i=0}^{k-1} \frac{\alpha^i}{i!} \right). \tag{3.13}$$

For $k = 1$, $\frac{\partial}{\partial \alpha}\left(\sum_{i=0}^{k-1} \frac{\alpha^i}{i!}\right) = 0$. And for $k > 1$, $\frac{\partial}{\partial \alpha}\left(\sum_{i=0}^{k-1} \frac{\alpha^i}{i!}\right) = \sum_{i=1}^{k-1} \frac{\alpha^{i-1}}{(i-1)!} = \sum_{i=0}^{k-2} \frac{\alpha^i}{i!}$. Therefore, we have $\frac{\partial P(X_\alpha \geq k)}{\partial \alpha} = e^{-\alpha} \frac{\alpha^{k-1}}{(k-1)!} > 0$, $\forall k \in \mathbb{Z}^+$ and $\alpha > 0$. With any given $\lambda > \lambda' > 0$, we can deduce that $P(X_\lambda \geq k) - P(X_{\lambda'} \geq k) = \int_{\lambda'}^{\lambda} \frac{\partial P(X_\alpha \geq k)}{\partial \alpha} d\alpha > 0$, $\forall k \in \mathbb{Z}^+$. □

This lemma is particularly singled out since it is used in many of the following propositions.

**Proposition 3.1** *The utility function $U(\lambda, w)$ monotonously increases with the type $\lambda$ and the quality $w$, where $\lambda > 0$ and $w \in \mathbb{N}$. In addition, the marginal increase of $U(\lambda, w)$ with respect to $w$ gets smaller as $w$ increases, as shown in Fig. 3.2.*

***Proof*** Consider a fixed value $w \in \mathbb{N}$ and $\lambda > \lambda' > 0$. If $w = 0$, we have $U(\lambda, w) = U(\lambda', w) = 0$ according to the definition. If $w > 0$, then $U(\lambda, w) - U(\lambda', w) =$

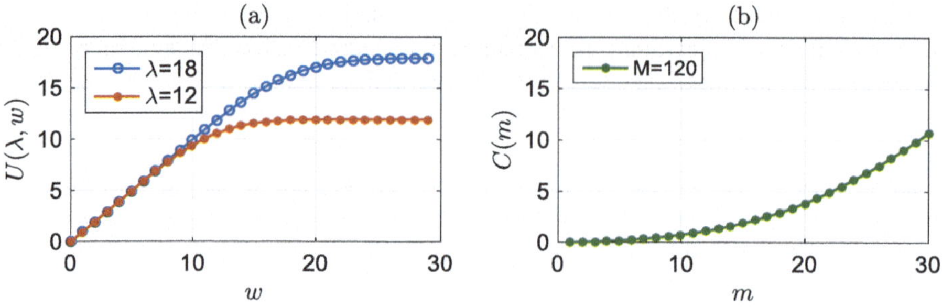


**Fig. 3.2** A simple illustration of the profiles of the utility function and the cost function. (**a**) UAV's utility function. (**b**) MBS's cost function

$\sum_{k=1}^{k=w} \big[P(X_\lambda \geq w) - P(X_{\lambda'} \geq w)\big] > 0$ according to Lemma 3.1. Therefore, $U(\lambda, w)$ monotonously increases with $\lambda$.

Now consider a fixed $\lambda > 0$ and $\forall w > w' \geq 0$, where $w, w' \in \mathbb{N}$. We have $U(\lambda, w) - U(\lambda, w') = P(X_\lambda \geq w) + \cdots + P(X_\lambda \geq w' + 1) \geq P(X_\lambda \geq w) > 0$. Therefore, $U(\lambda, w)$ monotonously increases with $w$.

For a fixed $\lambda > 0$ and $\forall w \geq 1$, we have $U'(w) = U(\lambda, w) - U(\lambda, w - 1) = P(X_\lambda \geq w)$. And for $w \geq 2$, we have $U''(w) = U'(w) - U'(w - 1) = -P(X_\lambda = w - 1) < 0$. Therefore, the marginal increase of $U(\lambda, w)$ with respect to $w$ gets smaller as $w$ increases. □

This proposition provides a basic property for us to design the optimal contract.

Based on Lemma 3.1 and Proposition 3.1, we exploit another important property of $U(\lambda, w)$, which says that a certain amount of quality improvement is more attractive to a higher type UAV than a lower type UAV. This property can be referred to as the "increasing preference (IP) property," and we write it as the following proposition:

**Proposition 3.2 (IP Property)** *For any UAV types $\lambda > \lambda' > 0$ and channel qualities $w > w' \geq 0$, the following inequality holds: $U(\lambda, w) - U(\lambda, w') > U(\lambda', w) - U(\lambda', w')$.*

***Proof*** According to the definitions of the utility function in (3.4) and (3.5), we have

$$U(\lambda, w) - U(\lambda, w') = P(X_\lambda \geq w) + \cdots + P(X_\lambda \geq w' + 1), \tag{3.14}$$

$$U(\lambda', w) - U(\lambda', w') = P(X_{\lambda'} \geq w) + \cdots + P(X_{\lambda'} \geq w' + 1). \tag{3.15}$$

Based on Lemma 3.1, each term in (3.14) is greater than each corresponding term in (3.15). Therefore we can obtain $U(\lambda, w) - U(\lambda, w') > U(\lambda', w) - U(\lambda', w')$. □

With the help of this property, we are able to deduce the best pricing strategy in the next subsubsection.

#### 3.1.2.2 Optimal Pricing Strategy

In this subsubsection, we use fixed quality assignment $\{w_t\}$ to analytically deduce the optimal pricing strategy $\{p_t\}$.

Based on the previous work on contract theory (such as in [28]), the IC and IR constraints and the IP property of the utility function in a contract design problem can directly lead to the conclusion as below:

**Proposition 3.3** *For the contract $\{(w_t, p_t)\}$ with the IC and IR constraints and the IP property, the following statements are simultaneously satisfied:*

- *The relation of types and qualities: $\lambda_i < \lambda_j \implies w_i \leq w_j$.*
- *The relation of qualities and prices: $w_i < w_j \iff p_i < p_j$.*

This conclusion contains basic properties of a feasible contract. It indicates that a higher price has to be associated with a higher quality, and a higher quality means a higher price should be charged. Although different qualities are not allowed to be associated with the same price, it is possible that different types of UAVs are assigned with the same quality and the same price.

**Lemma 3.2** *For the contract $\{(w_t, p_t)\}$ with the IC and IR constraints and the IP property, the following three conditions are the necessary conditions and sufficient conditions to determine a feasible pricing:*

- $0 \leq w_1 \leq w_2 \leq \cdots \leq w_T$,
- $0 \leq p_1 \leq U(\lambda_1, w_1)$,
- $p_{k-1} + A \leq p_k \leq p_{k-1} + B$, *for* $k = 2, 3, \cdots, T$, *where* $A = [U(\lambda_{k-1}, w_k) - U(\lambda_{k-1}, w_{k-1})]$ *and* $B = [U(\lambda_k, w_k) - U(\lambda_k, w_{k-1})]$.

***Proof***

**Necessity** These 3 conditions can be deduced from the IC and IR constraints and the IP property as follows: (1) Since $\{\lambda_1, \lambda_2, \cdots \lambda_T\}$ is written in the ascending order, we have $0 \leq w_1 \leq w_2 \leq \cdots \leq w_T$ and $0 \leq p_1 \leq p_2 \leq \cdots \leq p_T$ according to Proposition 3.3, where $w_i = w_{i+1}$ if and only if $p_i = i_{i+1}$. (2) Considering the IR constraint of $\lambda_1$-type UAVs, we can directly obtain $0 \leq p_1 \leq U(\lambda_1, w_1)$. Here, if $w_x = 0$, then $U(\lambda_t, w_t) = 0$ and $p_t = 0$ for any $t \leq x$. (3) Considering the IC constraint for the $k$-type and the $(k-1)$-type where $k > 1$, the corresponding expressions are given by $U(\lambda_k, w_k) - p_k \geq U(\lambda_k, w_{k-1}) - p_{k-1}$, and $U(\lambda_{k-1}, w_{k-1}) - p_{k-1} \geq U(\lambda_{k-1}, w_k) - p_k$. As we focus on the possible scope of $p_k$, we can deduce that $p_{k-1} + [U(\lambda_{k-1}, w_k) - U(\lambda_{k-1}, w_{k-1})] \leq p_k \leq p_{k-1} + [U(\lambda_k, w_k) - U(\lambda_k, w_{k-1})]$.

**Sufficiency** We have to prove that the prices $\{(p_t)\}$ determined by these conditions satisfy the IC and IR constraints. And the basic idea is to use mathematical induction, from $(w_1, p_1)$ to $(w_T, p_T)$, by adding the quality-price terms once at a time into the whole contract. For writing simplicity, the contract that only contains the first $k$ types of UAVs is denoted as $\Psi(k)$, where $\Psi(k) = \{(w_t, p_t)\}$, $1 \le t \le k$. First, we can verify that $w_1 \ge 0$ and $0 < p_i < U(\lambda_1, w_1)$ provided by the above conditions is feasible in $\Psi(1)$, since the IR constraint $U(\lambda_1, w_1) - p_i > 0$ is satisfied and the IC constraint is not useful in a single-type contract.

In the rest part of our proof, we show that if $\Psi(k)$ is feasible, then $\Psi(k+1)$ is also feasible, where $k+1 \le T$. To this end, we need to prove that (1) the newly added $\lambda_{k+1}$-type complies with its IC and IR constraints, given by

$$\begin{cases} U(\lambda_{k+1}, w_{k+1}) - p_{k+1} \ge U(\lambda_{k+1}, w_i) - p_i, & \forall i = 1, 2, \cdots, k, \\ U(\lambda_{k+1}, w_{k+1}) - p_{k+1} \ge 0, \end{cases} \tag{3.16}$$

and (2) the existing $k$ types still comply with their IC constraints with the addition of $\lambda_{k+1}$-type, given by

$$U(\lambda_i, w_i) - p_i \ge U(\lambda_i, w_{k+1}) - p_{k+1}, \quad \forall i = 1, 2, \cdots, k. \tag{3.17}$$

*First, we prove (3.16)*: Since $\Psi(k)$ is feasible, the IC constraint of $\lambda_k$-type should be satisfied, given by $U(\lambda_k, w_i) - p_i \le U(\lambda_k, w_k) - p_k, \forall i = 1, 2, \cdots, k$. Based on the right inequality in the third condition, we have $p_{k+1} \le p_k + U(\lambda_{k+1}, w_{k+1}) - U(\lambda_{k+1}, w_k)$. By adding up these two inequalities, we have $U(\lambda_k, w_i) - p_i + p_{k+1} \le U(\lambda_k, w_k) + U(\lambda_{k+1}, w_{k+1}) - U(\lambda_{k+1}, w_k), \forall i = 1, 2, \cdots, k$. According to the IP property, we can obtain that $U(\lambda_k, w_k) - U(\lambda_k, w_i) \le U(\lambda_{k+1}, w_k) - U(\lambda_{k+1}, w_i), \forall i = 1, 2, \cdots, k$, since $\lambda_{k+1} > \lambda_k$ and $w_k \ge w_i$. Again, by combining these two inequalities together, we can prove the IC constraint of the $\lambda_{k+1}$-type, given by $U(\lambda_{k+1}, w_{k+1}) - p_{k+1} \ge U(\lambda_{k+1}, w_i) - p_i, \forall i = 1, 2, \cdots, k$. The IR constraint of the $\lambda_{k+1}$-type can be easily deduced from the above IC constraint since $U(\lambda_{k+1}, w_i) - p_i \ge U(\lambda_i, w_i) - p_i \ge 0, \forall i = 1, 2, \cdots, k$. And therefore, we have $U(\lambda_{k+1}, w_{k+1}) - p_{k+1} \ge 0$.

*Then, we prove (3.17)*: Since $\Psi(k)$ is feasible, the IC constraint of $\lambda_i$-type, $i = 1, 2, \cdots, k$, should be satisfied, given by $U(\lambda_i, w_k) - p_k \le U(\lambda_i, w_i) - p_i, \forall i = 1, 2, \cdots, k$. Based on the left inequality in the third condition, we have $p_k + U(\lambda_k, w_{k+1}) - U(\lambda_k, w_k) \le p_{k+1}$. By adding up the above two inequalities, we have $U(\lambda_i, w_k) + U(\lambda_k, w_{k+1}) - U(\lambda_k, w_k) \le U(\lambda_i, w_i) - p_i + p_{k+1} \forall i = 1, 2, \cdots, k$. According to the IP property, we can obtain that $U(\lambda_i, w_{k+1}) - U(\lambda_i, w_k) \le U(\lambda_k, w_{k+1}) - U(\lambda_k, w_k), \forall i = 1, 2, \cdots, k$, since $\lambda_k \ge \lambda_i$ and $w_{k+1} \ge w_k$. Again, by combining the above two inequalities together, we can prove the IC constraint of the existing types, $\lambda_i, \forall i = 1, 2, \cdots, k$, given by $U(\lambda_i, w_i) - p_i \ge U(\lambda_i, w_{k+1}) - p_{k+1}$.

So far, we have proved that $\Psi(1)$ is feasible, and if $\Psi(k)$ is feasible then $\Psi(k+1)$ is also feasible. We can conclude that the final contract $\Psi(T)$ which includes all the types is feasible. Therefore, these three necessary conditions are also sufficient conditions. □

It provides an important guideline to design the prices for different types of UAVs. It implies that with fixed quality assignment $\{w_t\}$, the proper scope of the price $p_k$ depends on the value of $p_{k-1}$.

In the following, we provide the optimal pricing strategy of the MBS manager with fixed quality assignment $\{w_t\}$. Here we call $\{w_t\}$ a feasible quality assignment if $w_1 \leq w_2 \leq \cdots \leq w_T$ and $\sum_{t=1}^{T} w_t \leq M$, i.e., the first condition in Lemma 3.2 is satisfied and the channel number constraint is also satisfied. The maximum achievable revenue of the MBS manager with fixed and feasible quality assignment $\{w_t\}$ is given by

$$R^*(\{w_t\}) = \max_{\{p_t\}} \left[ \sum_{t=1}^{T} \left(N_t \cdot p_t\right) - c\left(\sum_{t=1}^{T} N_t \cdot w_t\right) \right]. \tag{3.18}$$

From the above equation we can see that, the key point is to maximize $\sum_{t=1}^{T} \left(N_t \cdot p_t\right)$, since the cost function is constant with fixed quality assignment $\{w_t\}$. Accordingly, we provide the following proposition for the optimal pricing strategy:

**Proposition 3.4 (Optimal Pricing Strategy)** *Given that* $\left\{(w_t, p_t)\right\}$ *is a feasible contract with feasible quality assignment* $\{w_t\}$*, the unique optimal pricing strategy* $\{\hat{p}_t\}$ *is*

$$\begin{cases} \hat{p}_1 = U(\lambda_1, w_1), \\ \hat{p}_k = \hat{p}_{k-1} + U(\lambda_k, w_k) - U(\lambda_k, w_{k-1}), \quad \forall k = 2, 3, \cdots, T. \end{cases} \tag{3.19}$$

***Proof*** By comparing (3.19) with Lemma 3.2, we can find that $\{\hat{p}_t\}$ is a feasible pricing strategy. In the following, we first prove that $\{\hat{p}_t\}$ is optimal, then prove that it is unique.

*Optimality* In the condition that quality assignment $\{w_t\}$ is fixed, $\{\hat{p}_t\}$ is optimal if and only if $\sum_{t=1}^{T} \left(N_t \cdot \hat{p}_t\right) \geq \sum_{t=1}^{T} \left(N_t \cdot p_t\right)$, where $\{p_t\}$ is any pricing strategy that satisfies the conditions in Lemma 3.2. Let us assume that there exists another better strategy $\{\tilde{p}_t\}$ for the MBS manager, i.e., $\sum_{t=1}^{T} \left(N_t \cdot \tilde{p}_t\right) \geq \sum_{t=1}^{T} \left(N_t \cdot \hat{p}_t\right)$. Since $N_t > 0$ for all $t = 1, 2, \cdots, T$, there is at least one $k \in \{1, 2, \cdots, T\}$ that satisfies $\tilde{p}_k > \hat{p}_k$. To guarantee that $\{\tilde{p}_t\}$ is still feasible, the following inequality must be complied according to Lemma 3.2: $\tilde{p}_k \leq \tilde{p}_{k-1} + U(\lambda_k, w_k) - U(\lambda_k, w_{k-1})$, if $k > 1$. Since $\tilde{p}_k > \hat{p}_k$, we have $\hat{p}_k < \tilde{p}_{k-1} + U(\lambda_k, w_k) - U(\lambda_k, w_{k-1})$, if $k > 1$. By substituting (3.19) into the above inequality, we have $\tilde{p}_{k-1} > \hat{p}_k + U(\lambda_k, w_k) - U(\lambda_k, w_{k-1}) = \hat{p}_{k-1}$, if $k > 1$. Repeat this process and we can finally obtain the result that $\tilde{p}_1 > \hat{p}_1 = U(\lambda_1, w_1)$, which contradicts with Lemma 3.2 where $p_1$ should not exceed $U(\lambda_1, w_1)$. Due to this contradiction, the above assumption that $\{\tilde{p}_t\}$ is better than $\{\hat{p}_t\}$ is impossible. Therefore, $\{\hat{p}_t\}$ is the optimal pricing strategy for the MBS manager.

*Uniqueness* Assume that there exists another pricing strategy $\{\tilde{p}_t\} \neq \{\hat{p}_t\}$, such that $\sum_{t=1}^{T}\left(N_t \cdot \tilde{p}_t\right) = \sum_{t=1}^{T}\left(N_t \cdot \hat{p}_t\right)$. Since $N_t > 0$ for all $t = 1, 2, \cdots, T$, there is at least one $k \in \{1, 2, \cdots, T\}$ that satisfies $\tilde{p}_k \neq \hat{p}_k$. If $\tilde{p}_k > \hat{p}_k$, then the same contradiction occurs just like we have discussed above. If $\tilde{p}_k < \hat{p}_k$, then there must exist another $\tilde{p}_l > \hat{p}_l$ to maintain $\sum_{t=1}^{T}\left(N_t \cdot \tilde{p}_t\right) = \sum_{t=1}^{T}\left(N_t \cdot \hat{p}_t\right)$. Either way, the contradiction is unavoidable, which implies that the optimal pricing strategy $\{\hat{p}_t\}$ is unique.

□

We write the general formula of the optimal prices $\{\hat{p}_t\}$ as

$$\hat{p}_t = U(\lambda_1, w_1) + \sum_{i=1}^{t} \theta_i, \qquad \forall t = 2, \cdots T, \tag{3.20}$$

where $\theta_1 = 1$ and $\theta_i = U(\lambda_i, w_i) - U(\lambda_i, w_{i-1})$ for $i = 2, \cdots T$. The optimal pricing strategy is able to maximize $R$ and achieve $R^*$ with any given feasible quality assignment. However, what $\{w_t\}$ is able to maximize $R^*$ and achieve the overall maximum value $\hat{R}$ is still unsolved.

### 3.1.2.3 Optimal Quality Assignment Problem

In this part, we analyze the optimal quality assignment problem based on the results in Sect. 3.1.2.2, and transform this problem into an easier form, as a preparation for the dynamic programming algorithm in Sect. 3.1.2.4.

The optimal quality assignment problem is given by

$$\begin{aligned} &\hat{R} = \max_{\{w_t\}} \left[R^*(\{w_t\})\right], \\ &s.t. \sum_{t=1}^{T} N_t w_t \leq M, \;\; w_1 \leq w_2 \leq \cdots \leq w_T, \;\; \text{and } w_t = 0, 1, 2 \cdots, \end{aligned} \tag{3.21}$$

where $R^*(\{w_t\})$ is the best revenue of a given quality assignment as given in (3.18). Based on the optimal pricing $\{\hat{p}_t\}$ in (3.20), we derive the expression of $R^*\left(\{w_t\}\right)$ as

$$R^*\left(\{w_t\}\right) = \sum_{t=1}^{T}\left[C_t \cdot U(\lambda_t, w_t) - D_t \cdot U(\lambda_{t+1}, w_t)\right] - C\left(\sum_{t=1}^{T} N_t \cdot w_t\right), \tag{3.22}$$

where $C_t = \left(\sum_{i=t}^{T} N_i\right)$, $D_t = \left(\sum_{i=t+1}^{T} N_i\right)$ for $t < T$, and $D_T = 0$. Here, we are able to guarantee that $C_t > D_t \geq 0, \forall t = 1, 2, \cdots, T$, since $N_t > 0, \forall t = 1, 2, \cdots, T$. We can observed from (3.22) that $w_i$ and $w_j$ $(i \neq j)$ are separated from each other in the first term. This is a non-negligible improvement to find the best $\{w_t\}$.

**Definition 3.1** A set of functions $\left\{G_t(w_t)\middle| t = 1, 2, \cdots T\right\}$, with the quality $w_t$ as the independent variable of $G_t(\cdot)$, with $C_t$ and $D_t$ ($C_t > D_t \geq 0$) as the constants of $G_t(\cdot)$, is given by

$$G_t(w_t) = C_t \cdot U(\lambda_t, w_t) - D_t \cdot U(\lambda_{t+1}, w_t), \quad w_t = 0, 1, 2, \cdots \quad \forall t = 1, 2, \cdots T. \tag{3.23}$$

Based on (3.22) and Definition 3.1, we have $R^*\big(\{w_t\}\big) = \sum_{t=1}^{T} G_t(w_t) - C\big(\sum_{t=1}^{T} N_t \cdot w_t\big)$. The meaning of $G_t(w_t)$ is the independent gain of setting $w_t$ for the $\lambda_t$-type UAVs regardless of the cost.

Based on $\{G_t(w_t)\}$, we can rewrite the optimization problem in (3.21) as

$$\begin{aligned}&\hat{R} = \max_{\{w_t\}} \left[\sum_{t=1}^{T} G_t(w_t) - C\left(\sum_{t=1}^{T} N_t w_t\right)\right]\\ &s.t. \sum_{t=1}^{T} N_t w_t \leq M, \;\; w_1 \leq w_2 \leq \cdots \leq w_T, \;\; \text{and } w_t = 0, 1, 2 \cdots\end{aligned} \tag{3.24}$$

This problem can be further transformed into an equivalent one, given by

$$\begin{aligned}&\hat{R} = \max_{\{W=0,1,\cdots,M\}} \left\{\max_{\{w_t\}} \left[\sum_{t=1}^{T} G_t(w_t)\right] - C\big(W\big)\right\},\\ &s.t. \sum_{t=1}^{T} N_t w_t \leq W, \;\; w_1 \leq w_2 \leq \cdots \leq w_T, \;\; \text{and } w_t = 0, 1, 2 \cdots\end{aligned} \tag{3.25}$$

where the original problem is divided into $M + 1$ subproblems (with different settings of $W$). Here, we have $W \in \mathbb{Z}$ and $W \in [0, M]$, which can be comprehended as the possible value of $\sum_{t=1}^{T} N_t w_t$. From this formulation, we can see that the overall optimal revenue can be acquired by comparing the best revenue of $M + 1$ subproblems. Since $C(W)$ is fixed in each subproblem, in the following we only focus on how to maximize $\sum_{t=1}^{T} G_t(w_t)$, given as

$$\begin{aligned}&\max_{\{w_t\}} \sum_{t=1}^{T} G_t(w_t),\\ &s.t. \sum_{t=1}^{T} N_t w_t \leq W, \;\; w_1 \leq w_2 \leq \cdots \leq w_T, \;\; \text{and } w_t = 0, 1, 2 \cdots\end{aligned} \tag{3.26}$$

By calculating all the best results of (3.26) with different possible values of $W$, we are able to obtain the optimal solution of (3.25) by further taking into account $C(W)$.

Therefore, we regard (3.26) as the key problem to be solved. The proposed dynamic programming algorithm for this problem is presented in the next subsubsection.

#### 3.1.2.4 Algorithm for the MBS Optimal Contract

In what follows, we first show the way of considering (3.26) as a distinctive form of the knapsack problem [29], then provide our recurrence formula to calculate its maximum value $G_{max}$, next present the method to find the parameters $\{w_t\}$ that achieve $G_{max}$, and finally provide an overview of whole solution including the optimal quality assignment $\{\hat{w}_t\}$ and the optimal pricing $\{\hat{p}_t\}$.

1. *A special knapsack problem*: First, we have to take a look at the constraints about the optimization parameters $\{w_i\}$. Since $w_i = 0, 1, 2 \cdots$ and $\sum_{t=1}^{T} N_t w_t \leq W$, we have $w_t \leq W$. To distinguish from the notation of *weight* in the following discussions, we use $K$ instead of $W$ as the common upper bound of $w_t$, $\forall t \in [1, T]$, where $K \leq W$. And we rewrite the constraint as $w_t \leq K$. Therefore for each $t$, there are totally $K+1$ optional values of $w_t$, given by $\{0, 1 \cdots K\}$. And the corresponding results of $G_t(w_t)$ are $\{G_t(0), G_t(1), G_t(2), \cdots, G_t(K)\}$, which represent the values of different object that we can choose. In addition, we interpret the constraint $\sum_{t=1}^{T} N_t w_t \leq W$ as the weight constraint in the knapsack problem, where $W$ is the weight capacity of the bag and setting $w_t = k$ means taking up the weight of $kN_t$.

For the convenience of understanding, we list the values and the weight of different options in Table 3.2. Each row presents all the options of a type and we should choose an option for each type. And the $k$th option in the $t$th row provides us with the value of $G_t(k)$ and the weight of $kN_t$. Due to the constraint of $w_1 \leq w_2 \leq \cdots \leq w_T$, we cannot choose the $(k+1)$th, $(k+2)$th $\cdots$ options in the $t$th row if we have already chosen the $k$th option in the $(t+1)$th row. Therefore, the algorithm introduced below is basically to start from the last row and end at the first row.

2. *The recurrence formula to calculate the maximum value $G_{max}$*: The key nature of designing a dynamic programming algorithm is to find the subproblems of the overall problem and write the correct recurrence formula. Here we define $OPT(t, k, w)$, $\forall t \in [1, T]$, $\forall k \in [0, K]$ and $\forall w \in [0, W]$, as the optimal outcome that includes the decisions from the $T$th row to the $t$th row, with the conditions that (1) the $k$th option in the $t$th row is chosen and (2) the occupied weight is no more than $w$. Since the algorithm starts from the $T$th row, we first provide the calculation of $OPT(T, k, w)$, $\forall k \in [0, K]$ and $\forall w \in [0, W]$, given as

**Table 3.2** All the optional objects to be selected

| | Type | Optional values | Corresponding weights |
|---|---|---|---|
| 1 | Type $\lambda_1$ | $G_1(0)$ $G_1(1)$ $G_1(2)$ $\cdots$ $G_1(K)$ | $0$ $N_1$ $2N_1$ $\cdots$ $KN_1$ |
| 2 | Type $\lambda_2$ | $G_1(0)$ $G_2(1)$ $G_2(2)$ $\cdots$ $G_2(K)$ | $0$ $N_2$ $2N_2$ $\cdots$ $KN_2$ |
| $\cdots$ | $\cdots$ | $\cdots$ | $\cdots$ |
| T | Type $\lambda_T$ | $G_T(0)$ $G_T(1)$ $G_T(2)$ $\cdots$ $G_T(K)$ | $0$ $N_T$ $2N_T$ $\cdots$ $KN_T$ |

$$OPT\big(T,k,w\big)=\begin{cases}G_T(k), & \text{if } w \geq kN_t,\\ -\infty, & \text{if } w < kN_t,\end{cases} \tag{3.27}$$

where $-\infty$ implies that $OPT(T, k, w)$ is impossible to be achieved due to the lack of weight capacity. This expression is straightforward since it only includes the $T$th row in Table 3.2. From $OPT\big(T, k, w\big)$, we can calculate $OPT\big(t, k, w\big)$ for all $t \in [1, T-1]$, $k \in [0, K]$ and $w \in [0, W]$ by the following recurrence formula:

$$OPT\big(t,k,w\big)=\begin{cases}\max\limits_{l=k,\cdots,K}\Big[G_t(k)+OPT\big(t+1,l,w-kN_t\big)\Big], & \text{if } w \geq kN_t,\\ -\infty, & \text{if } w < kN_t.\end{cases} \tag{3.28}$$

This formula implies that if we want to choose $k$ in the $t$th row, then the option that made in the $(t+1)$th row must be within $[k, K]$ due to the constraint of $w_1 \leq \cdots \leq w_T$. In addition, choosing $k$ in the $t$th row with total weight limit of $w$ indicates that there is only $w - kN_t$ left for the other rows from $t+1$ to $T$. And if $w - kN_t < 0$, the outcome is $-\infty$ since choosing $k$ in the $t$th row is impossible.

Let $G_{max}$ denote $\max\limits_{\{w_t\}} \sum\limits_{t=1}^{T} G_t(w_t)$, then we have the following expression:

$$G_{max} = \max_{k=0\cdots K}\Big[OPT\big(1,k,W\big)\Big]. \tag{3.29}$$

Thus we have to calculate $OPT(1, k, W)$ for all $k \in [0, K]$, by iteratively using (3.28).

3. *The method to find the parameters* $\{w_t\}$ *that achieve* $G_{max}$: Note that the above calculation only considers the value of the optimal result $G_{max}$. To record what exact values of $\{w_t\}$ are chosen for this optimal result by the algorithm, we have to add another data structure, given as $D(t, k, w)$. We let $D(t, k, w) = l$ if $OPT(t, k, w)$ chooses $l$ to maximize its value in the upper line of (3.28), which is given by

$$D\big(t,k,w\big)=\begin{cases}arg\max\limits_{l=k,\cdots,K}\Big[G_t(k)+OPT\big(t+1,l,w-kN_t\big)\Big], & \text{if } w \geq kN_t,\\ 0, & \text{if } w < kN_t.\end{cases} \tag{3.30}$$

After acquiring $G_{max}$ in (3.29), we can use $D(t, k, w)$ to inversely find the optimal values of $\{w_t\}$ along the "path" of the optimal solution. Specifically, we have

$$\begin{cases} \hat{w}_1 = arg \max_{k=0\cdots K} \Big[ OPT\big(1, k, W\big) \Big], \\ \hat{w}_t = D\big(t-1, \hat{w}_{t-1}, W - \sum_{i=1}^{t-2} \hat{w}_i N_i\big), \quad \forall t = 2, \cdots, T, \end{cases} \tag{3.31}$$

where we define $\sum_{i=1}^{t-2} \hat{w}_i N_i$ as 0 if $t-2 = 0$, just for writing simplicity.

4. *An overview of whole solution*: By now, we have presented the key part of our solution, i.e., the dynamic programming algorithm to solve the optimization problem in (3.26). The problem in (3.25), i.e., our final goal, can be directly solved by setting different values of $W$ in (3.26) and comparing the corresponding results with the consideration of $C(W)$.

A summary of our entire solution is given in Algorithm 1. It can be observed that the computational complexity of calculating $OPT(t, k, w)$ for all $k \in [0, K]$, $w \in [0, W]$ and $t \in [1, T]$ is $O(TK^2W)$. Therefore, the overall complexity is $O(MTK^2W)$, which can also be written as $O(TM^4)$ since $W \leq M$ and $K \leq W$. Although $M^4$ seems to be non-negligible, there are usually no more than hundreds of available channels of a MBS to be allocated in practice.[5]

---

**Algorithm 1:** The algorithm of optimal contract for UAV offloading

**Input**: Type information $\{\lambda_1, \cdots \lambda_T\}$, $\{N_1, \cdots N_T\}$, and the number of total channels $M$.
**Output**: Optimal pricing strategy $\{\hat{p}_1, \cdots \hat{p}_T\}$, optimal quality assignment $\{\hat{w}_1, \cdots \hat{w}_T\}$.

**begin**
  Calculate $G_t(k)$ for all $t \in [1, T]$ and $k \in [0, M]$ by (3.23);
  Calculate $C(m)$ for all $m \in [0, M]$ by (3.8);
  Initialize $\hat{R} = 0$, $w_t = 0$ for all $t \in [1, T]$, and $p_t = 0$ for all $t \in [1, T]$;
  **for** $W$ *is from* 0 *to* $M$ **do**
    Let $K = W$, to be the upper bound for each $w_i$;
    Calculate $OPT(T, k, w)$ for $\forall k \in [0, K]$ and $\forall w \in [0, W]$ by (3.27);
    Calculate $OPT(t, k, w)$ for $\forall k \in [0, K]$, $\forall w \in [0, W]$ and $\forall t \in [1, T-1]$ by (3.28);
    Acquire $G_{max}$ from $\{OPT(1, w, t)\}$ according to (3.29);
    **if** $G_{max} - C(W) > \hat{R}$ **then**
      Update the overall maximum revenue $\hat{R} = G_{max} - C(W)$;
      Update $\hat{w}_t$ for all $t \in [1, T]$ according to (3.30) and (3.31);
      Update $\hat{p}_t$ for all $t \in [1, T]$ based on $\{\hat{w}_t\}$ according to (3.19);
    **end**
  **end**
**end**

---

[5] In addition, by deleting unnecessary values in Table 3.2, we can resolve a whole contract with $M = 300$ in only 1 s.

#### 3.1.2.5 Socially Optimal Contract

To better discuss the effectiveness of the above MBS optimal contract, in the following, we briefly discuss another contract that aims to maximize social welfare. Before that, we briefly explain the true meaning of social welfare. In our context, social welfare indicates the sum of the revenue of the MBS and the total profits of the UAVs (as shown in Fig. 3.3), which also means the increase of the number of users that can be served by the overall system.[6] Therefore, social welfare can be seen as the parameter to indicate the effectiveness of the UAV offloading system.

The objective of socially optimal contract is given by

$$\hat{S} = \max_{\{w(\lambda)\},\{p(\lambda)\}} \sum_{\lambda \in \Lambda} \Big(N_\lambda \cdot U\big(\lambda, w(\lambda)\big)\Big) - C\Big(\sum_{\lambda \in \Lambda} N_\lambda \cdot w(\lambda)\Big), \tag{3.32}$$

where the first term is the total utility of the UAVs, the second term is the cost of the MBS, and we omit the constraints since they are the same with those in (3.12). This optimization problem has a similar structure with (3.12) and can be solved by the proposed dynamic programming algorithm with only minor changes. To calculate the optimal $\{w(\lambda)\}$ and $\{p(\lambda)\}$, we need to replace $G_t(k)$ by $N_t U(t, k)$ in line 2 of Table 3.2. In addition, we use $U_{max}$ to replace $G_{max}$ to represent the maximum overall utility of the UAVs. At last, the equation in line 11 of Table 3.2 should be replaced by $\hat{S} = U_{max} - C(W)$ to represent the maximum social welfare. For writing convenience, in the rest part of this section, we call the solutions of (3.12) and (3.32) as the "MBS optimal contract" and the "socially optimal contract," respectively. In addition, the relation of social welfare and MBS's revenue is illustrated in Fig. 3.3.

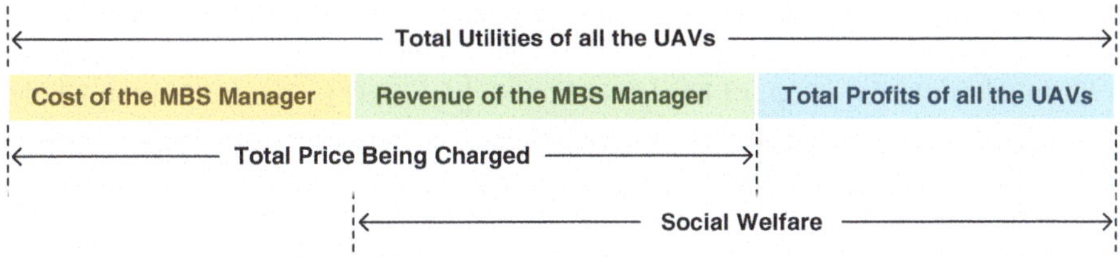


**Fig. 3.3** The relation of the social welfare, the revenue of the MBS manager, and the total profit of the UAVs operators

[6]Based on Fig. 3.3, although *Social Welfare* = *Revenue of the MBS Manager* + *Total Profits of all the UAVs*, we can also express it as *Social Welfare* = *Total Utilities of all the UAVs* − *Cost of the MBS Manager*, just as given in Eq. (3.32).

### 3.1.3 Theoretical Analysis and Discussions

In this subsection, we briefly discuss the impact of the height of the UAVs, $H$. Since $H$ influences the optimal revenue of the MBS $\hat{R}$, through the types of the UAVs $\{\lambda_t\}$, we first discuss the impact of $H$ on $\{\lambda_t\}$ in Sect. 3.1.3.1 and then discuss the impact of $\{\lambda_t\}$ on $\hat{R}$ in Sect. 3.1.3.2.

#### 3.1.3.1 The Impact of the Height on the UAV Types

We first define $\sigma_n = \varepsilon_n / S_n$, as the average density of active users in the effective coverage region of UAV$_n$. And we provide the following proposition:

**Proposition 3.5** *With fixed transmission power $P_{UAV}$ and $P_{MBS}$, fixed terrain parameters $a$, $b$, $\eta_{Los}$, and $\eta_{NLoS}$, fixed average active user density $\sigma_n$, fixed horizontal locations of the UAVs, and unified height $H \in [0, +\infty)$ of the UAVs, there exists a height $\hat{H}_n$ that can maximize the effective offloading area of UAV$_n$.*

***Proof*** We use $\phi$ (in radian) instead of $\theta$ (in degree) to denote the elevation angle, where $\phi = \theta \cdot \pi / 180°$. For a user with horizontal distance $r$ to the UAV, the average pathloss is given by $\overline{L}_{UAV}(\phi, r) = L_{LoS}(d) P_{LoS}(\theta) + L_{NLoS}(d)\big[1 - P_{LoS}(\theta)\big]$. With minor deduction, we have

$$\overline{L}_{UAV}(\phi, r) = L_{NLoS}(d) - \eta \cdot P_{LoS}(\theta), \tag{3.33}$$

where $\eta = \eta_{NLoS} - \eta_{LoS} < \eta_{NLoS}$, $d = \frac{r}{\cos\phi}$, $\theta = \frac{180°}{\pi}\phi$. By denoting $L_{NLoS}(d)$ as $L_1$ and denoting $\eta P_{LoS}(\theta)$ as $L_2$ to simplify the writing, we can provide the following assertions based on (3.1): As $\phi$ increases from 0 to $\pi/2$, $L_1$ increases monotonously from $L_{NLoS}(r)$ to infinity, while $L_2$ monotonously increases within a sub-interval of $(0, \eta)$. Therefore, $0 < \overline{L}_{UAV}(0, r) < L_{NLoS}(r)$, and $\overline{L}_{UAV}(\phi, r) \to +\infty$ as $\phi \to \pi/2$. In addition, $\overline{L}_{UAV}(\phi, r)$ has lower bound, $\big[L_{NLoS}(r) - \eta_{NLoS}\big]$, in the whole definition domain $[0, \pi/2]$. By considering the partial derivative of $\overline{L}_{UAV}(\phi, r)$ with respect to $\phi$, we have

$$\frac{\partial \overline{L}_{UAV}(\phi, r)}{\partial \phi} = \frac{\partial L_1}{\partial \phi} - \frac{\partial L_2}{\partial \phi} = \frac{20}{\ln 10}\tan\phi - \frac{180° \pi^{-1} a b \eta \exp[-b(\theta - a)]}{\{1 + a\exp[-b(\theta - a)]\}^2}. \tag{3.34}$$

where we have $\partial L_1 / \partial \phi = 0$ as $\phi = 0$, $\partial L_1 / \partial \phi \to +\infty$ as $\phi \to +\infty$, and $\partial L_2 / \partial \phi > 0$ as $\forall \phi \in [0, \pi/2)$. (Also note that Eq. (3.34) is the same as the one in [11].) Therefore, we can conclude that $\overline{L}_{UAV}(\phi, r)$ decreases near $\phi = 0$ and rapidly increases to $+\infty$ near $\phi = \pi/2$.

By now we have confirmed that: (a) $\overline{L}_{UAV}(\phi, r)$ decreases near $\pi = 0$; (b) $\overline{L}_{UAV}(\phi, r)$ increases to infinity as $\phi \to \pi/2$; and (c) $\overline{L}_{UAV}(\phi, r)$ has a lower bound in $[0, \pi/2)$. Therefore, there is at least one minimal value as $\phi \in (0, \pi/2)$

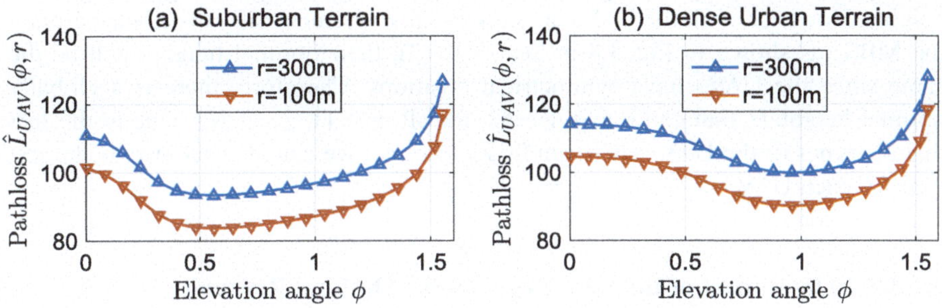


**Fig. 3.4** (**a**) Shows the pathloss in a typical suburban terrain, where parameters $a = 5$, $b = 0.2$, $\eta_{LoS} = 0.1$, and $\eta_{NLoS} = 21$. (**b**) Shows the pathloss in a typical dense urban terrain, where parameters $a = 14$, $b = 0.12$, $\eta_{LoS} = 1.6$, and $\eta_{NLoS} = 23$

that is smaller than $\overline{L}_{UAV}(0, r)$, which makes the existence of a minimum value as $\phi \in (0, \pi/2)$. Figure 3.4 provides an exemplary illustration of $\overline{L}_{UAV}(\phi, r)$ with different $r$ values.

The effective offloading region of the UAV, however, is based on the SNR of each possible location. Rigorous mathematical analysis would be highly difficult, thus only a simple discussion is provided as follows. Since we have assumed that the UAVs have the same height and the fixed horizontal locations, we can first conclude that, if a user is horizontally nearest to UAV $n$, then the SNR from UAV $n$ is always the largest among all the UAVs no matter how large $H$ is. Therefore, the user partition among UAVs is independent of $H$, and we only have to care about whether the SNR from UAV $n$ ($\gamma_{UAV_n}$) is greater than the SNR from the MBS ($\gamma_{MBS}$). For any given location, the scope of $H$ that satisfies $\gamma_{UAV_n} > \gamma_{MBS}$ can be either an empty interval or one or more disjoint intervals (called as the *effective height interval* of this user), depending on the number and the values of the minimal points of $\overline{L}_{UAV}(\phi, r)$.

At the height of $H$, the effective offloading area of UAV $n$ (given by $S_n$) depends on whether the value of $H$ resides in the effective height interval of each possible location on the ground. The theoretical deduction of the optimal height that maximizes $S_n$ is intractable. However, the existence of such optimal height can be guaranteed, since the effective height intervals are either empty or within $[0, +\infty)$. □

Since finding the optimal height is an intractable problem, the numerical method to obtain it can be done by numerically trying different values of $H$ in our algorithm and see which value achieves the highest revenue for the MBS operator, as shown in Fig. 3.8.

Note that this conclusion is different from existing studies (such as [30]), since we define "effective coverage region" of a UAV as the area that has a higher receive SNR from this UAV compared with the receive SNR from the MBS. From this proposition, we know that in the process of $H$ varying from 0 to $+\infty$, different UAVs are able to achieve their maximum effective offloading areas at different heights. However, if all the UAVs are horizontally symmetrically distributed around the MBS (as shown in Fig. 3.8 in Sect. 3.1.4.2), their optimal heights will be the same since the UAVs have symmetrical positions. Therefore, there is a globally optimal height $\hat{H}$ that can maximize $S_n$, for all $n \in \{1, 2, \cdots N\}$. Due to the fact that the types of the UAVs are given by $\lambda_n = \sigma_n S_n$, we can also achieve the largest type for each UAV.

#### 3.1.3.2 The Impact of the UAV Types on the Optimal Revenue

For any two random sets of types $\{\lambda_1, \cdots \lambda_{T_1}\}$ and $\{\lambda'_1, \cdots \lambda'_{T_2}\}$, there is no obvious relation of the outcomes of the corresponding two MBS optimal contracts. However, some properties can be explored when we add some constraints, as given in the following proposition:

**Proposition 3.6** *Given a fixed number of types $T$, two sets of types $\{\lambda_t\}$, $\{\lambda'_t\}$, and the constraint $\lambda_t \le \lambda'_t$, $\forall t \in [1, T]$, we have $\hat{R} \le \hat{R}'$, where $\hat{R}$ is the MBS's revenue of a MBS optimal contract with inputs $\{\lambda_t\}$ and $\hat{R}'$ is the MBS's revenue of a MBS optimal contract with inputs $\{\lambda'_t\}$.*

***Proof*** For the MBS optimal contract based on $\{\lambda_t\}$, the bandwidth allocation is denoted as $\{w_t\}$ and the corresponding cost of the MBS is denoted as $C\big(\sum w_t\big)$. If we change the types from $\{\lambda_t\}$ to $\{\lambda'_t\}$ and assume that the bandwidth allocation remains to be $\{w_t\}$, the cost of the MBS will still be $C\big(\sum w_t\big)$. Since $\lambda_t \le \lambda'_t$, we have $U(\lambda_t, w) < U(\lambda'_t, w)$ according to Proposition 3.1. And based on (3.19), we can deduce that $p_t$ will be greater, for any $t = 1 \cdots T$. Therefore, the sum of prices will get larger, and the revenue of the MBS will increase from $\hat{R}$ to $\hat{R}_w$. Note that the above discussion is based on the assumption that $\{w_t\}$ remain the same, which is probably not an optimal bandwidth allocation for $\{\lambda'_t\}$. If we run the algorithm in Sect. 3.1.2.4, the final revenue $\hat{R}'$ based on another bandwidth allocation $\{w'_t\}$ will be greater than $\hat{R}_w$. Therefore, we have $\hat{R} \le \hat{R}_w \le \hat{R}'$. □

With Propositions 3.5 and 3.6, we can directly obtain a conclusion that, there exists a highest value of the MBS's revenue by manipulating the height of the UAVs, as long as the UAVs are horizontally symmetrically distributed around the MBS, as shown in Sect. 3.1.4.

### *3.1.4 Simulation Results*

In this subsection, we simulate and compare the outcomes of the MBS optimal contract and the socially optimal contract under different settings. Simulation setups are given in Sect. 3.1.4.1; simulation results and corresponding discussions are provided in Sect. 3.1.4.2.

**Table 3.3** Simulation parameters

| | |
|---|---|
| Terrain parameters $a$ and $b$ | 11.95 and 0.136 |
| Additional pathloss parameters $\eta_{LoS}$ and $\eta_{NLoS}$ | 2 and 20 dB |
| Transmission power $P_{MBS}$ and $P_{UAV}$ | 10 W and 50 mW |
| Downlink transmission frequency $f$ | 3 GHz |
| Height of UAVs $H$ | Between 200 and 1000 m |
| Average active user density $\sigma_n$ (also as $\mu_n/S_n$) (km$^{-2}$) | Between 10 and 20 |
| Number of UAVs' types $T$ | Between 1 and 20 |
| Number of each type of UAVs $\{N_t\}$ | Between 1 and 10 |
| Average active user number of UAVs $\{\lambda_t\}$ | Between 1 and 20 |
| Average active user number of MBS $\lambda_{BS}$ | Between 10 and 200 |
| Number of total channels of MBS $M$ | Between 100 and 300 |
| Total energy of UAV$_n$ $E_n$ | Between 2000 and 8000 mAh |
| Stabilization energy consumption of UAV$_n$ during one time slot $e_n$ | 200 mAh |
| Moving energy consumption of UAV$_n$ between time slots $q_n$ | Between 1 and 5 mAh/m |

#### 3.1.4.1 Simulation Setups

We set $M$ within [100, 300], which is sufficient to generally evaluate a real system such as LTE [31]. The terrain parameters are set as $a = 11.95$ and $b = 0.136$, indicating a typical urban environment. We also set the transmission power as $P_{UAV} < P_{MBS}$, due to the typical consideration of UAVs that they have limited battery capacities. Details of the settings of all the parameters can be found in Table 3.3.

In the following simulations, we first study the UAV offloading system **based on the given UAV types** (i.e., fixed active user number for each UAV), from which we can acquire basic comprehension of the MSB optimal contract and the socially optimal contract, shown in Figs. 3.5, 3.6, and 3.7. Then we further study a more practical scenario where the height of the UAVs **determines the types of them**, shown in Fig. 3.8. At last we present the results of **multiple time slots** where the mobility and the energy constraint of a UAV influence its operator's long-term profit, shown in Fig. 3.9.

#### 3.1.4.2 Simulation Results and Discussions

We first illustrate the **typical structure of the contract** designed according to our algorithm, as given in Fig. 3.5, where we set $T = 10$, $\{N_t\} = (1, 1, \cdots 1)$, $\{\lambda_t\} = (1, 2, \cdots 10)$, and $M = 200$. All the four subplots show the patterns of $\{w_t\}$, $\{p_t\}$, and $\{U(t, w_t) - p_t\}$ with respect to different type $\lambda_t$. To be specific, subplots (a) and (b) show the results of lightly loaded MBS ($\lambda_{BS} = 120$) while (c) and (d) show the

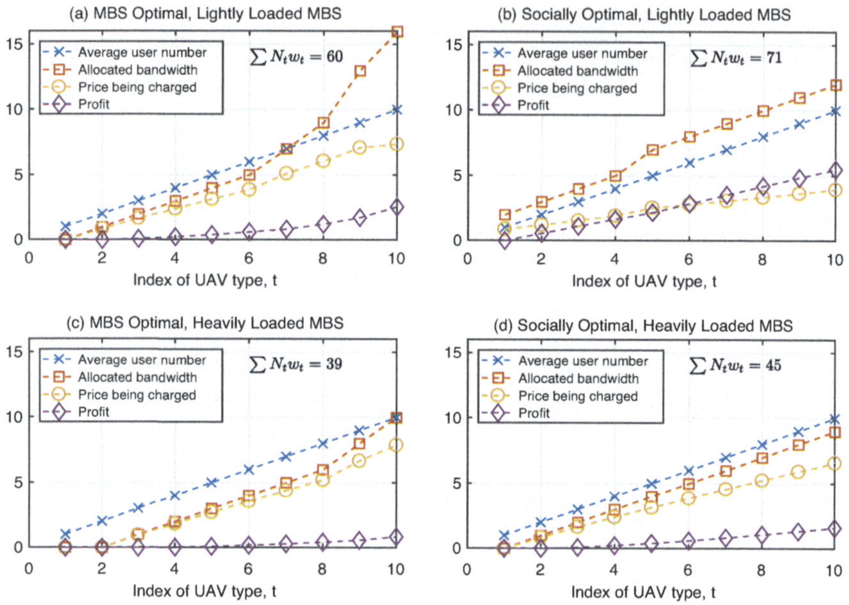


**Fig. 3.5** The structure of the optimal contracts where $T = 10$, $\{N_t\} = (1, 1, \cdots 1)$, $\{\lambda_t\} = (1, 2, \cdots 10)$, and $M = 200$, with $\lambda_{BS} = 120$ for (**a**) and (**b**), and $\lambda_{BS} = 160$ for (**c**) and (**d**). In addition, (**a**) and (**c**) show MBS optimal contracts while (**b**) and (**d**) show socially optimal contracts

results of heavily loaded MBS ($\lambda_{BS} = 160$). In addition, subplots (a) and (c) are the outcomes of MBS optimal contracts while (b) and (d) are the outcomes of socially optimal contracts. In any one of these subplots, we can see that a higher type of UAV is allocated with more channels but also a higher price. It can also be observed that a higher type gains more profit compared with a lower type, i.e., $U(i, w_i) - p_i \leq U(j, w_j) - p_j$ as long as $i < j$. In Fig. 3.5a, it is noticeable that for $\lambda_8$, $\lambda_9$ and $\lambda_{10}$-types, the allocated channels exceed their respective average user numbers. Such phenomenon is quite reasonable since a UAV needs more channels $w$ than its average active user number $\lambda$ to deal with the situation of burst access. And due to the IP property, higher types consider additional channels more valuable than lower types. Therefore, only $\lambda_8$, $\lambda_9$ and $\lambda_{10}$-types are allocated with excessive channels. By comparing Fig. 3.5a with b, or Fig. 3.5c with d, we find that a socially optimal contract allocates more channels than a MBS optimal contract, where we have 60 against 71 in (a) and (b), and 39 against 45 in (c) and (d). It can be considered that a socially optimal contract is more "generous" than a MBS optimal contract. By comparing Fig. 3.5a with c, or Fig. 3.5b with d, we can also find the difference of the numbers of totally allocated channels. This is because the cost of a heavily loaded MBS allocating the same number of channels is greater than that of a lightly loaded MBS.

To better explain the aforementioned **bandwidth differences**, we provide Fig. 3.6 to show how social welfare and the MBS's revenue change during the

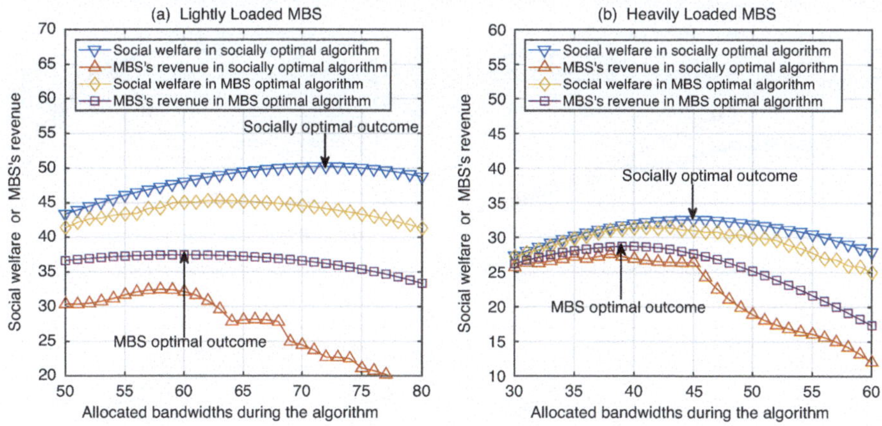


**Fig. 3.6** The change of social welfare and MBS's revenue during the socially optimal algorithm and MBS optimal algorithm, where $T = 10$, $\{N_t\} = (1, 1, \cdots 1)$, $\{\lambda_t\} = (1, 2, \cdots 10)$, $M = 200$, with $\lambda_{BS} = 120$ for (**a**) and $\lambda_{BS} = 160$ for (**b**)

algorithm with $W$ setting from 0 to $M$ (as described in line 5 in Table 3.2). In Fig. 3.6a, the upmost blue curve shows the change of social welfare during the socially optimal algorithm. The highest point of this curve represents the corresponding socially optimal contract, which makes $W = 71$ just as given in Fig. 3.5b. The lowermost orange curve shows the corresponding change of the MBS's revenue during the socially optimal algorithm. For the MBS optimal algorithm, the resulting curve of the MBS's revenue lies above the orange one from the socially optimal algorithm, while the resulting curve of social welfare lies below the blue one from the socially optimal algorithm. Since the two groups of curves do not coincide, we can deduce that the structure of the solutions of the two algorithms is not identical. For a fixed $W$, the MBS optimal algorithm somehow changes the allocation of channels among different types to increase the MBS's revenue, which results in a reduction of social welfare. And the bandwidth allocation of the MBS optimal contract is $W = 60$, just as given in Fig. 3.5a. In Fig. 3.6b, we also show the situation of heavily loaded MBS, where the relation of these curves is similar, as well as the reason that causes this.

Figure 3.7 illustrates **the impacts of the load of the MBS,** $\lambda_{BS}$, on the different part of the utility of the whole system as presented in Fig. 3.3. From Fig. 3.7a we can see that, the difference of allocated channels between the MBS optimal contract and the socially optimal contract becomes smaller as the load of MBS gets heavier. This is due to the fact that the cost of MBS rises fast when it is heavily loaded and neither the MBS optimal nor the socially optimal contract can allocate enough channels as desired. Figure 3.7b shows us that the MBS optimal contract is able to guarantee a high level of total prices being charged as the MBS is not heavily loaded. In addition, the total prices being charged according to the socially optimal contract is not monotonous and may rapidly change. For the case $\lambda_{BS} > 150$, although the total

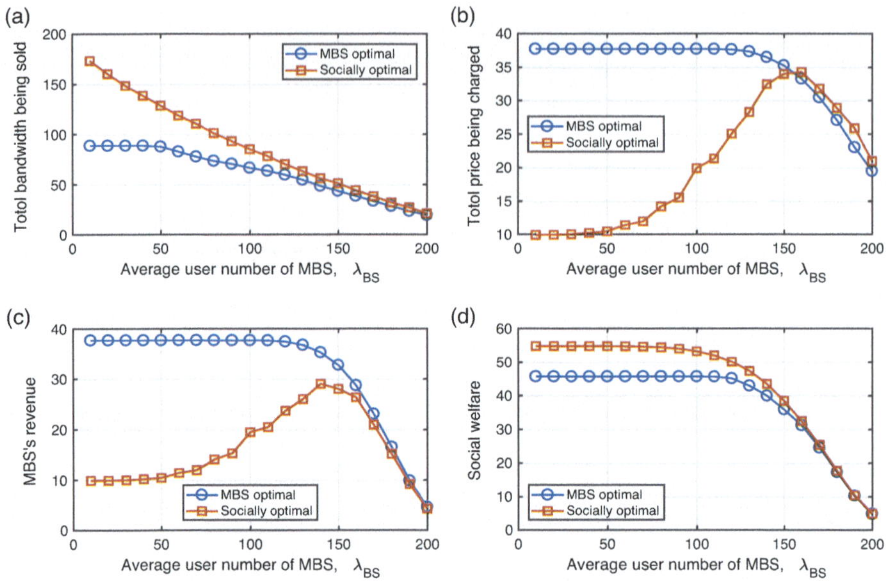


**Fig. 3.7** The impacts of $\lambda_{BS}$, where $T = 10$, $\{N_t\} = (1, 1, \cdots 1)$, $\{\lambda_t\} = (1, 2, \cdots 10)$ and $M = 200$

price being charged in the MBS optimal contract is lower than that in the socially optimal contract, the final revenue of the MBS is still higher in the MBS optimal contract as shown in Fig. 3.7c. This is because the MBS optimal contract has less total bandwidth being sold, which reduces the cost of the MBS. The social welfare is given in Fig. 3.7d, which implies that for both MBS and socially optimal contracts, a heavier loaded MBS could bring a lower overall system efficiency.

Then, we study **the impact of the height of the UAVs**, as presented in Fig. 3.8, where $M = 200$, $\lambda_{BS} = 150$. The considered 10 UAVs are located 1000 m horizontally from the MBS and symmetrically distributed. The average active user density of the effective offloading region of UAV$_n$ (i.e., $\sigma_n$) is set from $10\,\text{km}^{-2}$ to $20\,\text{km}^{-2}$. From the top three subplots in Fig. 3.8, we can see that the offloading regions of these UAVs first expand then shrink when the height of the UAVs monotonously increases. The maximum offloading areas can be achieved at $H = 674$, where the UAVs can cover the largest number of active users, as given in Fig. 3.8d. In addition, the MBS's revenue can be maximized when offloading areas become the largest, as discussed in Sect. 3.1.3. It can also be observed in Fig. 3.8f that the profile of the social welfare in the MBS optimal contract is very close to that of that social welfare in the socially optimal contract. In addition, the best height for the socially optimal contract ($H = 676$) is very close to the best height for the MBS optimal contract ($H = 674$). Therefore, we can infer that the height $H$ designated by the selfish MBS manager will generally keep a high social welfare. In other words, the performance of the overall system will not be significantly impaired.

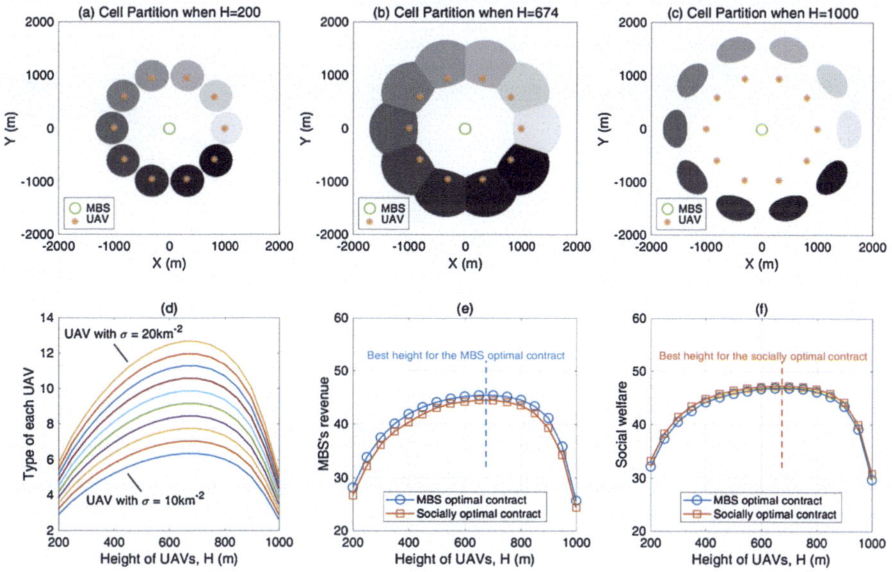


**Fig. 3.8** The impact of the height of UAVs. The subplots (**a**), (**b**), and (**c**) show the top views of the cell partition with different height settings. The white areas represent MBS's effective service regions, while gray areas represent UAVs' effective offloading regions. The subplot (**d**) provides the impact on the type of each UAV with different active user density. The subplots (**e**) and (**f**) illustrate the impacts of the height of UAVs on "MBS's revenue" and "social welfare," respectively

At last, we take a look at the influence of **UAV's mobility and energy constraint**. We generate the initial distribution of users according to Poisson point process (PPP), then acquire the distribution of users in the next time slot according to random walk with a maximum moving distance of 10 m. Ten UAV operators are added into the system with disjoint target region of users. Each UAV has a fixed cost of deploying, given by $C_n = 40$, $n = 1, 2, \cdots 10$. We focus on only one of these UAVs, which can adjust its horizontal location between time slots with a maximum moving distance of 20 m, to greedily maximize its number of covered users (based on the algorithm in [16]). Figure 3.9 shows the result, where the comparison of "fixed UAV" and "mobile UAV" is provided. Here we further set the mobility cost as 1 and 5 mAh/m to illustrate the difference between a low-cost movement and a high-cost movement. Since an adjustable UAV is able to cover as many users as possible in each time slot, the UAV's profit is expected to be higher. However, the energy consumption of mobility may also reduce the number of time slots of the deployment. Therefore, a low additional energy consumption ($q = 1$ mAh/m) of mobility could result in a better outcome for the UAV operator, while a high additional energy consumption ($q = 1$ mAh/m) of mobility could make it worse to adjust the position of UAV. Moreover, it can also be observed that the profit of the UAV operator has an approximate linear relation with the total energy of its UAV, since a higher battery capacity can

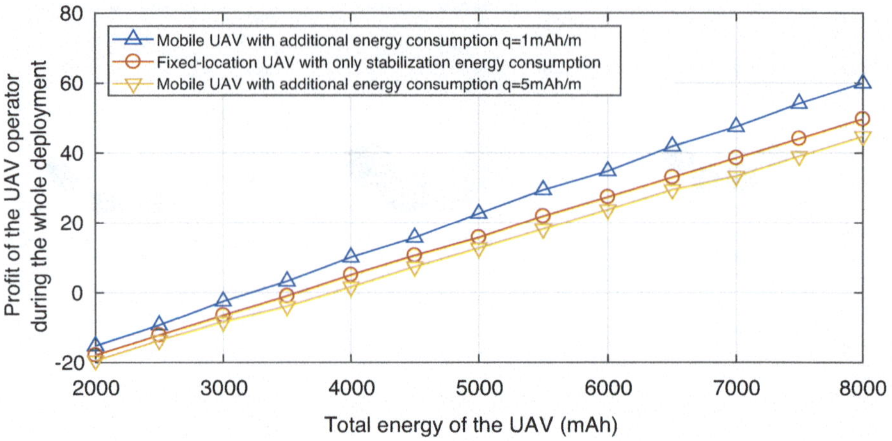


**Fig. 3.9** The impact of the mobility and battery capacity of the UAV, with $H = 200$ m, $M = 200$, $\lambda_{BS} = 120$

increase the time of deployment. To guarantee the total profit to be positive in the long term, the UAV operator should use a high-capacity battery for UAV offloading.

### *3.1.5 Summary*

In this section, we have focused on the scenario where the UAVs are deployed in a cellar network to better serve local mobile users. Considering the selfish MBS manager and the selfish UAV operators, we have modeled the utilities and the costs of spectrum trading among them and have formulated the problem of designing the optimal contract for the MBS manager. To deduce the optimal contract, we first have derived the optimal pricing strategy based on a fixed quality assignment, and then have analyzed and transformed the optimal quality assignment problem, in which way it can be solved by the proposed dynamic programming algorithm in polynomial time. In the simulations, by comparing with the socially optimal contract, we have found that the MBS optimal contract allocated fewer channels to the UAVs to guarantee a lower level of costs. In addition, the best height of the UAVs for the selfish MBS manager can keep a high performance of the overall system. Moreover, UAV's mobility is able to increase the long-term profit of the UAV operator, but a high-capacity battery is also necessary.

## 3.2 UAVs Serving as Relays

Recently, UAVs become especially helpful in the situations with widely scattered users, large obstacles such as hills or buildings that deteriorates the quality of links, and communication disabilities due to natural disasters [32]. Wireless communication with the assist of UAV, i.e., UAV relay, has been widely discussed. In UAV-aided relay networks, UAVs are deployed to provide wireless connectivities between two or more distant users or user groups without reliable direct communication links [10]. In [33], an energy efficiency maximization algorithm is proposed for UAV relay with circular trajectory. In [6], the authors study throughput maximization of a rectilinear trajectory UAV relay network. However, most of the works only consider the location of UAV as a fixed point or on a fixed trajectory. In practice, UAVs can move freely in the three-dimensional space to achieve a better performance, but the trajectory design and power control on this condition have not been well studied.

In this section, we consider a half-duplex uplink UAV relay network with a UAV, a BS, and a mobile device (MD). The UAV works as an amplify-and-forward (AF) relay, which is capable to adjust its transmit power and flying trajectory. We formulate the trajectory design and power control as a non-convex outage probability minimization problem. The problem is decoupled into trajectory design and power control subproblems. For these two subproblems, we address them by gradient descent method and extremum principles, respectively. Finally, an approximate solution of trajectory design and power control is obtained to approach the minimum outage probability.

The rest of this section is organized as follows. In Sect. 3.2.1, we describe the system model and formulate the outage probability minimization problem. In Sect. 3.2.2, a joint trajectory design and power control algorithm is given to solve this problem. Simulation results are presented in Sect. 3.2.3, and finally we summarize this section in Sect. 3.2.4.

### *3.2.1 System Model and Problem Formulation*

As shown in Fig. 3.10, we consider an uplink scenario in a cellular network with one BS and one MD which is beyond the coverage of BS [34]. A UAV works as an AF relay[7] to provide communication service for the MD. During the transmission, the UAV adjusts its location to improve the service quality. We assume that the transmission process contains $N$ time slots. In two consecutive time slots, the MD transmits signals to UAV in the first time slot, and the UAV amplifies and forwards the received signals from the MD to BS in the second time slot. We denote the locations of BS and MD by $B$ and $M$, respectively.

[7]The same method can also be applied to the UAV relay with the DF protocol as shown in [39].

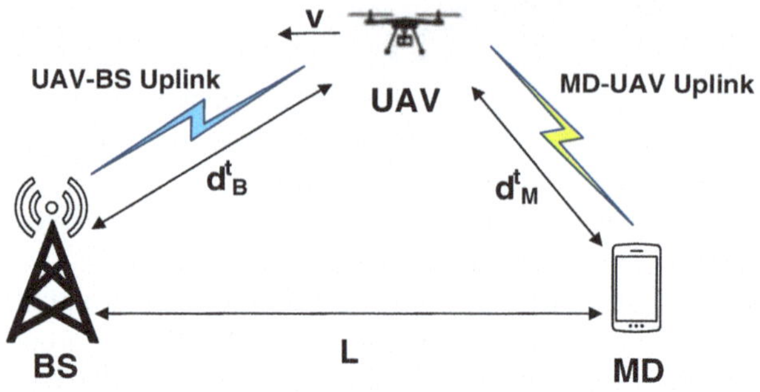


**Fig. 3.10** System model for communication with UAV relay

Let $L$ be the distance between the BS and MD. In time slot $t$, the distance between MD and UAV and the distance between UAV and BS are given by $d_M^t$ and $d_B^t$, respectively. The flying distance of UAV from time slot $i$ to time slot $j$ is $d_{i,j}$. We assume that the UAV can fly for a maximum distance of $v$ in each time slot, where $v \ll L$. The transmit power of MD in time slot $t$ is given by $P_M^t$, and the transmit power of UAV in time slot $(t+1)$ is given by $P_U^{t+1}$. The total transmit power in time slot $t$ and $(t+1)$ is constrained,[8] i.e., $P_M^t + P_U^{t+1} \leq P_{max}$.

In time slot $t$, the received power at UAV from the MD is expressed as [36]

$$P_{M,U}^t = P_M^t (d_M^t)^{-\alpha} \|h_M^t\|^2. \tag{3.35}$$

In time slot $(t+1)$, the received power at BS is shown as

$$P_{U,B}^{t+1} = P_U^{t+1} (d_B^{t+1})^{-\alpha} \|h_B^{t+1}\|^2, \tag{3.36}$$

where $\alpha$ is the pathloss exponent, and $h_M^t$, $h_B^{t+1}$ are independent small-scale channel fading coefficients with zero mean and unit variance. The noise at each node satisfies the Gaussian distribution with zero mean and $N_0$ as variance. The received signal of UAV relay is expressed as

$$Y_{UAV}^t = \sqrt{P_M^t (d_M^t)^{-\alpha}} h_M^t X_M^t + n_U^t, \tag{3.37}$$

where $X_M^t$ is the signal of unit energy from the MD, and $n_U^t$ is the noise received at the UAV relay. The amplification coefficient of the UAV relay is given by

$$G_t = \sqrt{P_U^{t+1} / \left(P_M^t (d_M^t)^{-\alpha} \|h_M^t\|^2 + N_0\right)}. \tag{3.38}$$

[8] With total power constraint, the optimal power solution that minimizes the outage probability can be obtained with various MD-UAV and UAV-BS distances ratios. It also guarantees that the maximum power efficiency can be reached with different transmission distances scenarios [35].

After being amplified by the UAV relay, the received signal at BS can be expressed as

$$\begin{aligned} Y_{BS}^{t+1} &= G_t\sqrt{P_M^t(d_M^t)^{-\alpha}(d_B^{t+1})^{-\alpha}}h_M^t h_B^{t+1}X_M^t \\ &+n_U^t G_t\sqrt{(d_B^{t+1})^{-\alpha}}h_B^{t+1}+n_B^{t+1}, \end{aligned} \tag{3.39}$$

where $n_B^{t+1}$ is the noise received at BS. According to (3.39), the SNR of the uplink network is given by

$$\gamma_t = \frac{P_M^t(d_M^t)^{-\alpha}\|h_M^t\|^2 G_t^2(d_B^{t+1})^{-\alpha}\|h_B^{t+1}\|^2}{N_0 G_t^2(d_B^{t+1})^{-\alpha}\|h_B^{t+1}\|^2+N_0}. \tag{3.40}$$

The outage probability is defined as the probability that the SNR falls below a predetermined threshold $\gamma_{th}$. Thus, the outage probability of the uplink can be derived by integrating the probability density function (PDF) of $\gamma_t$, shown as

$$P_{out}^t = P[\gamma_t \le \gamma_{th}] = \int_0^{\gamma_{th}} f(\gamma_t)d\gamma_t. \tag{3.41}$$

**Theorem 3.1** *The approximate expression of the outage probability in time slot t and* $(t+1)$ *is*

$$\begin{aligned} P_{out}^t &= 1-\exp(-\tfrac{N_0\gamma_{th}}{P_M^t(d_M^t)^{-\alpha}})\times(1+2V^2\ln V), \\ V &= \sqrt{(N_0\gamma_{th})/(P_U^{t+1}(d_B^{t+1})^{-\alpha})}. \end{aligned} \tag{3.42}$$

***Proof*** According to (3.40), we rewrite SNR $\gamma_t$ as

$$\gamma_t = \left(aP_{M,U}^t P_{U,B}^{t+1}\right)/\left(aP_{U,B}^{t+1}N_0+N_0\right), \tag{3.43}$$

where $a = G_t^2/P_U^{t+1}$. Variables $P_{M,U}^t$ and $P_{U,B}^{t+1}$ obey exponential distribution for their physical significance. The outage probability in (3.41) can be simplified as

$$P_{out}^t = P(P_{M,U}^t \le S+W), \tag{3.44}$$

where $S = N_0\gamma_{th}$ and $W = N_0\gamma_{th}/aP_{U,B}^{t+1}$. Let $\Phi = P_M^t(d_M^t)^{-\alpha}$, the outage probability can be rewritten by

$$\begin{aligned} P_{out}^t &= E_{S+W}\{P(P_{M,U}^t \le s+w|s+w)\} \\ &= E_{S+W}\{\textstyle\int_0^{s+w}(1/\Phi)\exp(-x/\Phi)dx\} \\ &= E_{S+W}\{1-\exp(-(s+w)/\Phi)\}. \end{aligned} \tag{3.45}$$

As $S$ and $W$ are independent variables, we further obtain

$$P_{out}^{t} = 1 - E_S\{\exp(-S/\Phi)\} \times E_W\{\exp(-W/\Phi)\}. \tag{3.46}$$

Since variable $S$ is a constant, $E_S$ can be expressed by

$$E_S\{\exp(-S/\Phi)\} = \exp(-(N_0\gamma_{th})/\Phi). \tag{3.47}$$

We can also derive the PDF of $W$ from the PDF of $P_{U,B}^{t+1}$, which is given by

$$\begin{aligned} f_W(w) &= \tfrac{d}{dw}P(W \le w) = \tfrac{d}{dw}P(\tfrac{N_0\gamma_{th}}{ay} \le w) \\ &= \tfrac{N_0\gamma_{th}}{aw^2} \times \tfrac{1}{\Psi}\exp(-\tfrac{N_0\gamma_{th}}{a\Psi w}), \end{aligned} \tag{3.48}$$

where $\Psi = P_U^{t+1}(d_B^{t+1})^{-\alpha}$. Thus, $E_W$ can be expressed as

$$\begin{aligned} &E_W\{\exp(-\tfrac{W}{\Phi})\} = \tfrac{1}{\Psi}\times \\ &\int_0^{+\infty}\exp(-\tfrac{w}{\Phi})\exp\left(-\tfrac{N_0\gamma_{th}}{aYw}\right)\times\tfrac{N_0\gamma_{th}}{aw^2}dw. \end{aligned} \tag{3.49}$$

By substituting (3.47) and (3.49) into (3.46), we have

$$\begin{aligned} &P_{out}^{t} = 1 - \tfrac{1}{\Psi}\exp(-\tfrac{N_0\gamma_{th}}{\Phi})\times \\ &\int_0^{+\infty}\exp(-\tfrac{w}{\Phi})\exp(-\tfrac{N_0\gamma_{th}}{a\Psi w})\times\tfrac{N_0\gamma_{th}}{aw^2}dw. \end{aligned} \tag{3.50}$$

According to the results in [37], (3.50) can be rewritten as

$$\begin{aligned} P_{out}^{t} &= 1 - \exp(-\tfrac{N_0\gamma_{th}}{\Phi}) \times 2VK_{-1}(2V), \\ V &= \sqrt{(N_0\gamma_{th}(\Phi + N_0))/(\Phi\Psi)}, \end{aligned} \tag{3.51}$$

where $K_{-1}(x)$ is the negative first order modified Bessel function of the second kind. Since $\Phi \gg N_0$, $V$ in (3.51) can be simplified as $V = \sqrt{\frac{N_0\gamma_{th}}{\Phi}}$. Note that the modified Bessel function of the second kind has the property $K_{-1}(x) = K_1(x)$. Thus, we expand $K_1(x)$ according to [37] and have

$$K_1(x) \simeq 1/x + x/2 \times \ln(x/2). \tag{3.52}$$

By substituting (3.52) into (3.51), the analytical approximate solution of the outage probability is shown as (3.42). □

Our objective is to minimize the outage probability by optimizing both the UAV trajectory and the power of UAV and MD. The expression of (3.42) shows that the outage probability in time slot $t$ is only affected by the power and location parameters in time slot $t$ and $(t+1)$. Therefore, we simplify the optimization objective as the outage probability in a single time slot, and the problem can be formulated by

$$\min_{P_M^t, P_U^{t+1}, d_M^t, d_B^{t+1}} \quad P_{out}^t, \tag{3.53a}$$

$$s.t. P_U^{t+1} + P_M^t \leq P_{max}, \tag{3.53b}$$

$$P_U^{t+1} \geq 0,\ P_M^t \geq 0, \tag{3.53c}$$

$$d_{t,t+1} \leq v, \tag{3.53d}$$

where (3.53b) and (3.53c) are the power constraints for the UAV and the MD, and (3.53d) shows the UAV mobility constraint.

### 3.2.2 Power and Trajectory Optimization

The expression of (3.42) shows that the joint power and trajectory optimization problem (4.7) is non-convex. In this section, we tackle the problem through alternating minimization, where trajectory design and power control are optimized iteratively.

The algorithm is illustrated in Algorithm 2. In each iteration, we design the trajectory given the power control results obtained by the last iteration, and then solve the power control subproblem given the UAV trajectory. In iteration $k$, let $S^k = \sum_{t=1}^{N} P_{out}^t$, the algorithm converges when $S^k - S^{k-1} < \epsilon$, where $\epsilon$ is a predefined error tolerance threshold.

---

**Algorithm 2:** Power and trajectory optimization algorithm

---

**Input**: Total transmit power $P_{max}$, and the maximum speed $v$.
**Output**: Transmit powers $P_M^t$, $P_U^{t+1}\}$, and location parameters $d_M^t$, $d_B^{t+1}$.

```
begin
    Initialize k = 0, S^0 = 0, P_M^t = P_U^{t+1} = P_max/2, ∀t = 1, 3, ··· , N;
    repeat
        k = k + 1;
        for t is from 1 to N do
            Solve trajectory design subproblem (3.54) for slot t;
        end
        for t is from 1 to N do
            Solve power control subproblem (3.57) for slot t;
        end
    until S^k − S^{k−1} ≤ ε;
end
```

---

#### 3.2.2.1 Trajectory Design

Given the power control variables $P_M^t$ and $P_U^{t+1}$, (4.7) can be expressed as

$$\min_{d_M^t, d_B^{t+1}} P_{out}^t, \tag{3.54a}$$

$$\boldsymbol{s.t.} d_{t,t+1} \leq v. \tag{3.54b}$$

Problem (3.54) is also non-convex. To achieve a local minimum outage probability, the UAV will fly in the direction with the maximum outage probability descent velocity, i.e., $-\nabla P_{out}^t$. Let $(0, 0, 0)$ and $(L, 0, 0)$ be the locations of the BS and the MD, respectively. In time slot $t$, we denote the location of UAV by $\boldsymbol{l_t} = (x_t, y_t, z_t)$, and the trajectory by $\boldsymbol{\Delta l_t} = (\Delta x_t, \Delta y_t, \Delta z_t)$, with $|\boldsymbol{\Delta l_t}| \ll |\boldsymbol{l_t}|$. When the high-order terms are neglected, the trajectory direction of the UAV, i.e., the gradient of the outage probability function is expressed as

$$\begin{aligned} -\nabla P_{out}^t &= ((R-M)x_t + ML)\hat{\boldsymbol{i}} + (R-M)y_t\hat{\boldsymbol{j}} \\ &\quad +(R-M)z_t\hat{\boldsymbol{k}}, \end{aligned} \tag{3.55}$$

where

$$\begin{aligned} M &= \tfrac{N_0\gamma_{th}}{P_M^t}\left((x_t-L)^2 + y_t^2 + z_t^2\right)^{\alpha/2-1}(1+Q\ln Q), \\ R &= \tfrac{N_0\gamma_{th}}{P_U^{t+1}}(x_t^2 + y_t^2 + z_t^2)^{\alpha/2-1}(1+\ln Q), \\ Q &= \tfrac{N_0\gamma_{th}}{P_U^{t+1}}(x_t^2 + y_t^2 + z_t^2)^{\alpha/2}; \end{aligned} \tag{3.56}$$

and $\hat{\boldsymbol{i}}$, $\hat{\boldsymbol{j}}$, and $\hat{\boldsymbol{k}}$ are the unit vectors of $x$, $y$, and $z$ axis, respectively.

Since $v \ll L$, we set the step size of the gradient descent process as $v$ for each time slot. We also set a minimum outage probability threshold $\delta$, with $\delta \to 0^-$. When $-\nabla P_{out}^t \geq \delta$, it is regarded that the minimum outage probability is achieved, and the UAV stops moving. In the trajectory design for time slot $(t+1)$, the location of UAV is updated as $\boldsymbol{l_{t+1}} = \boldsymbol{l_t} + \boldsymbol{\Delta l_t}$.

*Remark 3.1* When $N$ is sufficiently large, the outage probability will tend to be stable at the minimum value.

**Theorem 3.2** *When the outage probability is minimized, BS, MD, and UAV are collinear, and $d_B^t < d_M^t$ is satisfied.*

***Proof*** When $-\nabla P_{out}^t = 0$, the outage probability is minimized. In (3.55), it can be easily found that the root of $-\nabla P_{out}^t = 0$ contains $y_t = 0$ and $z_t = 0$, which means BS, MD, and UAV are collinear. The solution of $x_t$ satisfies $\frac{L}{x_t} = 1 - \frac{N}{M}$, which cannot be solved easily. When we substitute $x_t = L/2$, we have $\frac{L}{x_t} < 1 - \frac{N}{M}$. It can be proved that the right side of the inequation is monotonically increasing with $x_t$ at $x_t = L/2$ while the left side is monotonically decreasing. Therefore, the solution of $x_t$ exists in $0 < x_t < L/2$, showing that the UAV is closer to BS than to the MD. □

#### 3.2.2.2 Power Control

Given the UAV trajectory $\boldsymbol{l}_t$, problem (4.7) can be rewritten as

$$\min_{P_M^t, P_U^{t+1}} P_{out}^t, \tag{3.57a}$$

$$s.t. P_U^{t+1} + P_M^t \leq P_{max}, \tag{3.57b}$$

$$P_U^{t+1} \geq 0, P_M^t \geq 0. \tag{3.57c}$$

**Theorem 3.3** *The minimized outage probability is obtained when* $P_U^{t+1} + P_M^t = P_{max}$ *is satisfied.*

***Proof*** As shown in (3.50), both $P_{M,U}^t$ and $P_{U,B}^{t+1}$ are negatively related with $P_{out}^t$. Since the received power is positively related with the transmit power, the increment of the transmit power decreases the outage probability. Therefore, the minimum outage probability requires to maximize the total transmit power, i.e., $P_U^{t+1} + P_M^t = P_{max}$. □

We substitute $P_M^t = P_{max} - P_U^{t+1}$ into (3.57), and $P_{out}^t$ is a function of $P_U^{t+1}$. The outage probability function in (3.57) is convex with respect to $P_U^{t+1}$. Therefore, when $\frac{d(P_{out})}{dP_U^{t+1}} = 0$, the optimal power control is realized. The closed-form solution of power control is given in the following theorem.

**Theorem 3.4** *The proposed power control is mainly determined by* $d_M^t/d_B^{t+1}$. *When* $d_M^t \ll d_B^{t+1}$ *or* $d_M^t \gg d_B^{t+1}$, $P_U^{t+1} \to 0$, *and the transmit power of MD is* $P_M^t \to P_{max}$.

***Proof*** We substitute (3.42) into $\frac{d(P_{out})}{dP_U^{t+1}} = 0$ and assume that $P_{M,U}^t \gg N_0$. We then have

$$\ln\left(\frac{P_U^{t+1}}{N_0\gamma_{th}(d_B^{t+1})^\alpha}\right) = \frac{(P_U^{t+1})^2(d_M^t)^\alpha}{(P_{max} - P_U^{t+1})^2(d_B^{t+1})^\alpha} + 1. \tag{3.58}$$

With $\theta = \frac{P_U^{t+1}}{P_{max}}$ and $u = \frac{\theta}{1-\theta}$, Eq. (3.58) can be simplified as

$$u^2 = \ln(B/A) + \ln\theta, \tag{3.59}$$

where $A = \exp \frac{(d_M^t)^\alpha}{(d_B^{t+1})^\alpha}$, and $B = \frac{P_{max}}{eN_0\gamma_{th}(d_B^{t+1})^\alpha}$. Using Taylor expansion, $\ln(\theta) = \theta - 1 + o(\theta) \simeq -\frac{1}{u+1}$, and $u$ is the root of the following equation:

$$u^3 + u^2 - u \ln(B/A) + 1 - \ln(B/A) = 0. \tag{3.60}$$

It can be solved that

$$u = \left(-q/2 + \sqrt{q^2/4 + p^3/27}\right)^{1/3} + \left(-q/2 + \sqrt{q^2/4 - p^3/27}\right)^{1/3}, \tag{3.61}$$

where $p = -\ln(B/A) - 1/3$, and $q = 2 + 9\ln(B/A) + 27(1 - \ln(B/A))^2/27$. The approximate solution for the transmit power is $P_M^t = \frac{P_{max}}{u}$, and $P_U^{t+1} = P_{max} - P_M^t$. The power control solution is determined by the ratio of $A$ and $B$, which is mostly affected by $d_M^t$ and $d_B^{t+1}$. The key factor of power control is $d_M^t/d_B^{t+1}$, since $A$ is an exponential function of $d_M^t$ and $d_B^{t+1}$.

When $d_M^t \ll d_B^{t+1}$, we have $A \simeq 1$, and $B \gg A$; therefore, $|\ln \frac{B}{A}| \gg 1$. When $d_M^t \gg d_B^{t+1}$, it is shown that $A \gg B$ because $A$ is an exponential function of $d_M^t$, and we also have $|\ln \frac{B}{A}| \gg 1$. In both cases, (3.60) can be simplified as $u = 1$. The power control is given as $P_M^t = P_{max}$, and $P_U^{t+1} = 0$. It means that the relay will be redundant if it is too close to the source or destination. □

### 3.2.3 Simulation Results

In this subsection, we evaluate the performance of Algorithm 9. The selection of the simulation parameters is based on the existing works and 3GPP specifications [31, 38]. We consider $N = 1500$ time slots, and the distance between the MD and BS $L = 500$. The maximum initial MD-UAV and UAV-BS distances ($d_M^1$ and $d_U^1$) are set as $\frac{3}{5}L$. We set the maximum moving distance for UAV in each time slot as $v = 0.1$, and also set $\delta = -10^{-2}$, $\epsilon = -10^{-2}$. The maximum total transmit power $P_{max}$ is given as 26 dBm, and the noise variance $N_0$ is given as −96 dBm. The SNR threshold $\gamma_{th}$ is 0 dB, and the pathloss exponent $\alpha$ is 4. All curves are generated by averaging over $10^5$ instances.

We provide two schemes in comparison with the proposed power control and trajectory design scheme (PP-PT): proposed power control with circle trajectory scheme (PP-CT), and equal power allocation with fixed relay scheme (EP-FR). In the PP-CT, the trajectory is a circle whose center is $(L/2, 0, 0)$ and radius is 100. The initial location of UAV is a random point on this circle and the moving distance is $v$ for a time slot. The power control is the same as our proposed algorithm. In EP-FR, the location of UAV is fixed in different time slots, which

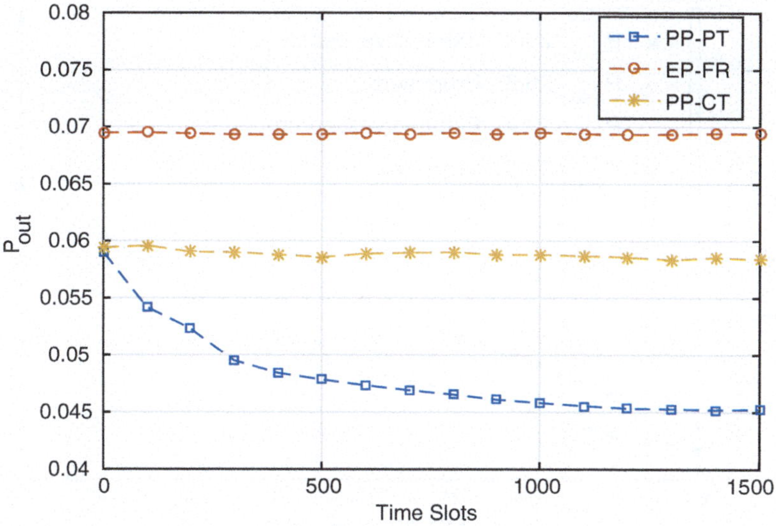


**Fig. 3.11** Time slot vs. outage probability

obeys uniform distribution in a circle area with $(L/2, 0, 0)$ being the center, and $L/2$ being the radius. The MD and the UAV use the same transmit power in different time slots.

Figure 3.11 depicts the average outage probability with time axis. The outage probability obtained by PP-PT decreases with time at the beginning. After about 1000 time slots, the outage probability decreases about 23% and turns stable at the minimum level, which is consistent with Remark 1. The average outage probability of PP-CT scheme is similar with PP-PT at the beginning, but it does not decrease with time since the trajectory is fixed. The outage probability obtained by EP-FR is about 18% higher than the scheme with PP-CT.

Figure 3.12 illustrates the distance between the MD and BS versus the minimum achievable outage probability. The number of transmission time slots $N$ is sufficiently large for the UAV to achieve the minimum outage probability with our proposed solution. The minimum outage probability of exhaustive search is obtained by enumerating the outage probability of over 1000 power control strategies and over 5000 UAV location possibilities. It is shown that the minimum outage probability increases monotonically with the increment of $L$. When the maximum total transmit power is raised from 26 dBm to 29 dBm, the outage probability will decrease about 40%. The difference between the minimum outage probability obtained by the proposed solution and the exhaustive search is less than 5% in all of our simulations.

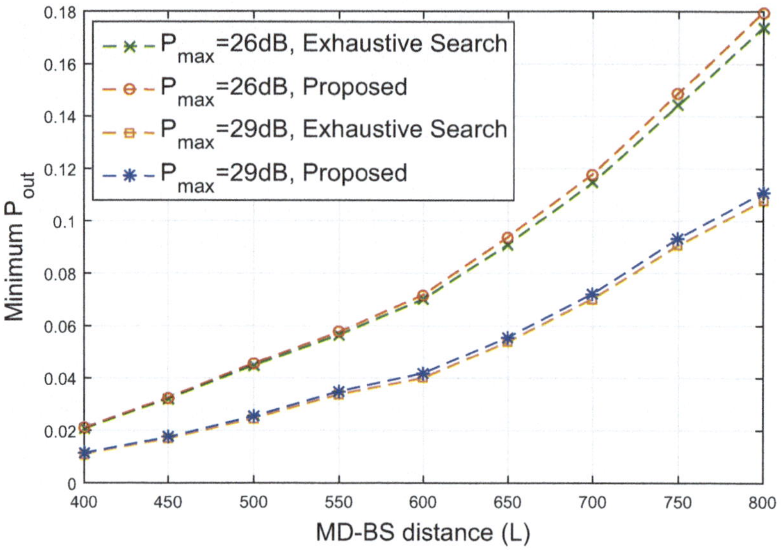


**Fig. 3.12** MD-BS distance (L) vs. minimum outage probability

### *3.2.4 Summary*

In this section, we have considered an uplink network model with a UAV working as an AF relay. An analytical expression of the outage probability has been derived. With power and UAV speed constraints, we have given a joint solution for the trajectory design and power control. The outage probability of proposed solution has converged to the minimum level, and outperformed the fixed power and circle trajectory schemes significantly. The minimum outage probability obtained by the proposed algorithm has been proved to be close to the minimum outage probability using the exhaustive search, with a difference less than 5%.

## References


1. S. Hayat, E. Yanmaz, R. Muzaffar, Survey on unmanned aerial vehicle networks for civil applications: a communications viewpoint. IEEE Commun. Surv. Tutorials **18**(4), 2624–2661 (2016)
2. N.R. Kuntz, Y.O. Paul, Development of autonomous cargo transport for an unmanned aerial vehicle using visual servoing. ASME Dyn. Syst. Control Conf. **7503**(39), 731–738 (2008)
3. P.M. Olsson, J. Kvarnströ, P. Doherty, O. Burdakov, K. Holmberg, Generating UAV communication networks for monitoring and surveillance, in *Proceedings of International Conference on Control Automation Robotics & Vision*, Singapore (2010)
4. E.W. Frew, T.X. Brown, Airborne communication networks for small unmanned aircraft systems. Proc. IEEE **96**(12), 2008–2027 (2008)

5. İ. Bekmezci, O.K. Sahingoz, Ş. Temel, Flying Ad-Hoc networks (FANETs): a survey. Ad Hoc Netw. **11**(3), 1254–1270 (2013)
6. Y. Zeng, R. Zhang, T.J. Lim, Throughput maximization for UAV-enabled mobile relaying systems. IEEE Trans. Commun. **64**(12), 4983–4996 (2016)
7. R. Amorim, H. Nguyen, P. Mogensen, I. Kovács, J. Wigard, T.B. Sørensen, Radio channel modelling for UAV communication over cellular networks. IEEE Wireless Commun. Lett. **6**(4), 514–517 (2017)
8. D.W. Matolak, R. Sun, Unmanned aircraft systems: air-ground channel characterization for future applications. IEEE Veh. Technol. Mag. **10**(2), 79–85 (2015)
9. S. Karapantazis, F. Pavlidou, Broadband communications via high-altitude platforms: a survey. IEEE Commun. Surv. Tutorials **7**(1), 2–31 (2005)
10. Y. Zeng, R. Zhang, T.J. Lim, Wireless communications with unmanned aerial vehicles: opportunities and challenges. IEEE Commun. Mag. **54**(5), 36–42 (2016)
11. A. Al-Hourani, S. Kandeepan, S. Lardner, Optimal LAP altitude for maximum coverage. IEEE Wireless Commun. Lett. **3**(6), 569–572 (2014)
12. M. Mozaffari, W. Saad, M. Bennis, M. Debbah, Drone small cells in the clouds: design, deployment and performance analysis, in *Proceedings of IEEE Global Communications Conference*, San Diego (2015)
13. M. Alzenad, A. El-Keyi, F. Lagum, H. Yanikomeroglu, 3D placement of an unmanned aerial vehicle base station (UAV-BS) for energy-efficient maximal coverage. IEEE Wireless Commun. Lett. **6**(4), 434–437 (2017)
14. M. Mozaffari, W. Saad, M. Bennis, M. Debbah, Unmanned aerial vehicle with underlaid device-to-device communications: performance and tradeoffs. IEEE Trans. Wirel. Commun. **15**(6), 3949–3963 (2016)
15. V.V.C. Ravi, H.S. Dhillon, Downlink coverage probability in a finite network of unmanned aerial vehicle (UAV) base stations, in *Proceedings of IEEE International Workshop on Signal Processing Advances in Wireless Communications*, Edinburgh (2016)
16. J. Lyu, Y. Zeng, R. Zhang, T.J.Lim, Placement optimization of UAV-mounted mobile base stations. IEEE Commun. Lett. **21**(3), 604–607 (2017)
17. M. Mozaffari, W. Saad, M. Bennis, M. Debbah, Efficient deployment of multiple unmanned aerial vehicles for optimal wireless coverage. IEEE Commun. Lett. **20**(8), 1647–1650 (2016)
18. M. Mozaffari, W. Saad, M. Bennis, M. Debbah, Optimal transport theory for power-efficient deployment of unmanned aerial vehicles, in *Proceedings of IEEE International Conference on Communications*, Kuala Lumpur (2016)
19. S. Rohde, C. Wietfeld, Interference aware positioning of aerial relays for cell overload and outage compensation, in *Proceedings of IEEE Vehicular Technology Conference*, Quebec City (2012)
20. A. Merwaday, İ. Güvenç, UAV assisted heterogeneous networks for public safety communications, in *Proceedings of IEEE Wireless Communications and Networking Conference Workshops*, New Orleans (2015), pp. 329–334
21. V. Sharma, M. Bennis, R. Kumar, UAV-assisted heterogeneous networks for capacity enhancement. IEEE Commun. Lett. **20**(6), 1207–1210 (2016)
22. M. Mozaffari, W. Saad, M. Bennis, M. Debbah, Optimal transport theory for cell association in UAV-enabled cellular networks. *IEEE Commun. Lett.* **21**(9), 2053–2056 (2016)
23. S. Jangsher, V.O.K. Li, Resource allocation in moving small cell network. IEEE Trans. Wirel. Commun. **15**(7), 4559–4570 (2016)
24. C.A. Wargo, G.C. Church, J. Glaneueski, M. Strout, Unmanned aircraft systems (UAS) research and future analysis, in *Proceedings of IEEE Aerospace Conference*, Big Sky (2014)
25. A. Valcarce, T. Rasheed, K. Gomez, S. Kandeepan, L. Reynaud, R. Hermenier, A. Munari, M. Mohorcic, M. Smolnikar, I. Bucaille, *Airborne Base Stations for Emergency and Temporary Events*, vol. 6. (Springer, Berlin, 2013), pp. 49–58
26. P. Bolton, M. Dewatripont, *Contract Theory* (MIT Press, London, 2005)
27. Z. Hu, Z. Zheng, L. Song, T. Wang, X. Li, UAV offloading: spectrum trading contract design for UAV-assisted cellular networks. IEEE Trans. Wirel. Commun. **17**(9), 6093–6107 (2018)

28. L. Gao, X. Wang, Y. Xu, Q. Zhang, Spectrum trading in cognitive radio networks: a contract-theoretic modeling approach. IEEE J. Sel. Areas Commun. **29**(4), 843–855 (2011)
29. S. Martello, P. Toth, *Knapsack Problems: Algorithms and Computer Implementations* (John & Sons, Inc., New York, 1990)
30. M. Alzenad, A. El-Keyi, H. Yanikomeroglu, 3d Placement of an unmanned aerial vehicle base station for maximum coverage of users with different QoS requirements. IEEE Wireless Commun. Lett. **7**(1), 38–41 (2018)
31. *Evolved Universal Terrestrial Radio Access (EUTRA) Physical Layer Procedures Release 12*, document 3GPP TS 36.213 (2014)
32. M. Erdelj, E. Natalizio, K.R. Chowdhury, I.F. Akyildiz, Help from the sky: leveraging UAVs for disaster management. *IEEE Pervasive Comput.* **16**(1), 24–32 (2017)
33. D. Choi, S. Kim, D. Sung, Energy-efficient maneuvering and communication of a single UAV-based relay. IEEE Trans. Aerosp. Electron. Syst. **50**(3), 2320–2327 (2014)
34. S. Zhang, H. Zhang, Q. He, K. Bian, L. Song, Power and trajectory optimization for UAV relay networks. IEEE Commun. Lett. **22**(1), 161–164 (2018)
35. S. Salari, M.Z. Amirani, I. Kim, D.I. Kim, J. Yang, Distributed beamforming in two-way relay networks with interference and imperfect CSI. IEEE Trans. Wirel. Commun. **15**(6), 4455–4469 (2016)
36. S. Zhang, B. Di, L. Song, Y. Li, Sub-channel and power allocation for non-orthogonal multiple access relay networks with amplify-and-forward protocol. IEEE Trans. Wirel. Commun. **16**(4), 2249–2261 (2017)
37. I.S. Gradshteyxn, I.M. Ryzhik, *Table of Integrals, Series, and Products* (Academic Press, San Diego, 2007)
38. H. Zhang, Y. Liao, L. Song, D2D-U: device-to-device communications in unlicensed bands for 5G system. IEEE Trans. Wirel. Commun. **16**(6), 3507–3519 (2017)
39. S. Zeng, H. Zhang, K. Bian, L. Song, UAV relaying: power allocation and trajectory optimization using decode-and-forward protocol, in *Proceedings of the IEEE ICC Workshops*, Kansas City (2018)

# Chapter 4
# Cellular Assisted UAV Sensing

The other application is to exploit UAVs for sensing purposes due to its advantages of on-demand flexible deployment, larger service coverage compared with the conventional fixed sensor nodes, and ability to hover. Due to the limited computation capability of UAVs, the sensory data needs to be transmitted to the BS for real-time data processing. In this regard, the cellular networks are necessarily committed to support the data transmission for UAVs, which we refer to as *Cellular assisted UAV Sensing*. Nevertheless, to support *real-time* sensing streaming, it is desirable to design *joint sensing and communication* protocols, develop novel beamforming and estimation algorithms, and study efficient distributed resource optimization methods.

In this chapter, we first study the energy efficiency maximization problem when the UAV serve as an aerial user in Sect. 4.1, and then discuss the task completion time minimization problem in a cooperative cellular Internet of UAVs in Sect. 4.2. To improve the communication quality in the cellular Internet of UAVs, we propose UAV-to-everything (U2X) communications in Sect. 4.3. In Sect. 4.4, we investigate the decentralized trajectory design problem for cellular Internet of UAVs. Finally, we present an application of cellular Internet of UAVs for air quality index (AQI) monitoring.

## 4.1 Cellular Internet of UAVs

UAV is an emerging facility which has been effectively applied in military, public, and civil applications [1]. Among these applications, the use of UAV to perform data sensing has been of particular interest owing to its advantages of on-demand flexible deployment, larger service coverage compared with the conventional fixed sensor nodes, and additional design degrees of freedom by exploiting its high mobility [2–4]. Recently, UAVs with cameras or sensors have entered the daily lives



H. Zhang et al., *Unmanned Aerial Vehicle Applications over Cellular Networks for 5G and Beyond*, Wireless Networks, https://doi.org/10.1007/978-3-030-33039-2_4

to execute various sensing tasks, e.g., air quality index monitoring [5], autonomous target detection [6], precision agriculture [7], and water stress quantification [8].

In these UAV sensing applications, the sensory data collected in such tasks needs to be immediately transmitted to the BSs for further processing in the servers [9], thereby posing low latency requirement on the wireless network. In the UAV ad hoc network [2, 10], the sensory data is transmitted over the unlicensed band, which cannot guarantee the QoS requirements. Therefore, the terrestrial cellular networks are considered as a promising enabler for UAV sensing applications, which we refer to as the cellular Internet of UAVs.

However, in the cellular Internet of UAVs, the sensory data such as live video streaming and high-resolution images needs to be transmitted with limited UAV on-board battery, thereby posing high uplink rate and low energy consumption requirements. In this section, we study a simplified cellular UAV system in which one UAV serves as an aerial user. The UAV performs data sensing tasks and transmits the sensory data to the BS simultaneously. We aim to maximize the energy efficiency (EE) of the system while guaranteeing the velocity and rate constraints. To solve this problem efficiently, we decompose it into two subproblems, namely UAV sensing optimization and UAV transmission optimization subproblems. In the UAV sensing optimization subproblem, we optimize the trajectory and the speed by convex optimization and differential methods, respectively. In the UAV transmission optimization subproblem, we can obtain the optimal transmission power solution by the extremum principles.

The rest of this section is organized as follows. In Sect. 4.1.1, we describe the system model, and formulate the EE maximization problem in Sect. 4.1.2. In Sect. 4.1.3, an EE maximization algorithm is designed to solve this problem. Simulation results are presented in Sect. 4.1.4, and finally we summarize this section in Sect. 4.1.5.

## *4.1.1 System Model*

We consider a cellular Internet of UAVs as shown in Fig. 4.1, which consists of one BS and one UAV [9]. The UAV is required to execute $N$ sensing tasks in sequence within the cell coverage, denoted by $\mathcal{N} = \{1, 2, \ldots N\}$. The UAV performs each task in two steps: UAV sensing and UAV transmission, and thus, the UAV takes a number of iterations of sensing and transmission to complete the whole tasks.

### 4.1.1.1 UAV Sensing

In the UAV sensing, it is important to design the trajectory of the UAV along which it moves towards the locations of a sequence of tasks. The trajectory of the UAV is designed by the BS, and sent to the UAV over the control channel. Without loss of generality, we denote the location of the BS by $(0, 0, H)$, and the sensing area of task $n$ by a circle with $(x_n, y_n, 0)$ being the center and $r$ being the radius. In time slot

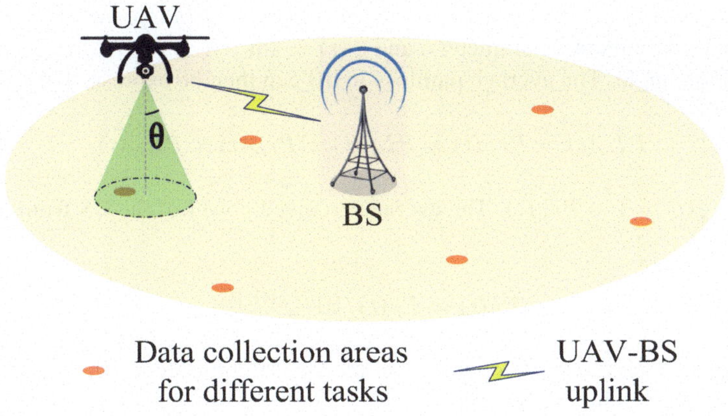


**Fig. 4.1** System model of a cellular Internet of UAVs

$t$, let $\boldsymbol{l}(t) = (x(t), y(t), z(t))$ be the location of UAV $i$, $\boldsymbol{v}(t) = (v_x(t), v_y(t), v_z(t))$ be its velocity, and $\boldsymbol{a}(t) = (a_x(t), a_y(t), a_z(t))$ be its accelerated velocity, with $\boldsymbol{v}(t) = \boldsymbol{l}(t) - \boldsymbol{l}(t-1)$, and $\boldsymbol{a}(t) = \boldsymbol{v}(t) - \boldsymbol{v}(t-1)$. Due to the space and mechanical limitations, the UAV has a maximum flight altitude $h_{max}$, a maximum velocity $v_{max}$, and a maximum accelerated velocity $a_{max}$.

To execute a sensing task, the UAV should cover the whole sensing area of the task. Define $\theta$ as the maximum angle that the UAV sensor can detect, and thus, the height of the UAV satisfies $z(t)\tan(\theta) \geq r$. The UAV at $\boldsymbol{l}(t) = (x(t), y(t), z(t))$ can perform the task whose center is within the area of

$$(x - x(t))^2 + (y - y(t))^2 \leq (z(t)\tan\theta - r)^2. \tag{4.1}$$

When performing a task, the UAV hovers for $t_0$ time slots to collect data with rate $R_s$.

#### 4.1.1.2 UAV Transmission

When performing data collection, the UAV uploads the collected sensory data to the BS concurrently. It is assumed that the UAV is assigned with a dedicated subchannel in the system, and thus, there is no interference in the UAV transmission process. For UAV transmission, we utilize the air-to-ground propagation model proposed in [11]. In time slot $t$, the line-of-sight (LoS) and non-line-of-sight (NLoS) pathloss models from the UAV to the BS are given by $PL_L(t) = L_{FS}(t) + 20\log(d_{UAV,BS}(t)) + \eta_{LoS}$, and $PL_N(t) = L_{FS}(t) + 20\log(d_{UAV,BS}(t)) + \eta_{NLoS}$, where $L_{FS}(t)$ is the free space pathloss given by $L_{FS}(t) = 20\log(f) + 20\log(\frac{4\pi}{c})$, $f$ is the system carrier frequency, and $d_{UAV,BS}(t)$ is the distance between the UAV and the BS. $\eta_{LoS}$ and $\eta_{NLoS}$ are additional attenuation factors due to the LoS and NLoS connections. Considering the antennas on the UAV and the BS placed vertically, the probability of LoS connection is given by $Pr_L(t) = (1 + \alpha \exp(-\beta(\phi(t) - \alpha)))^{-1}$, where $\alpha$ and $\beta$ are environmental parameters, and $\phi(t) = \sin^{-1}((z(t) - H)/d_{UAV,BS}(t))$ is the elevation angle. The average pathloss in dB can then be expressed as

$$PL_a(t) = Pr_L(t) \times PL_L(t) + Pr_N(t) \times PL_N(t), \tag{4.2}$$

where $Pr_N(t) = 1 - Pr_L(t)$. The average received power of the BS from the UAV is given by

$$P_R(t) = P_T(t)/10^{PL_a(t)/10}, \tag{4.3}$$

where $P_T(t)$ is the transmission power of the UAV in time slot $t$. The data rate from the UAV to the BS in time slot $t$ is

$$R(t) = W_B \times \log_2\left(1 + P_R(t)/\sigma^2\right), \tag{4.4}$$

where $W_B$ is the bandwidth of the subchannel, and $\sigma^2$ is the variance of AWGN with zero mean.

Let $\gamma(t)$ be a binary UAV sensing and transmission variable, where $\gamma(t) = 1$ if the UAV is performing sensing and data transmission in time slot $t$, otherwise, $\gamma(t) = 0$. We also assume that the UAV needs to finish the data transmission process concurrently with the data sensing process. Therefore, the transmission rate of the UAV should be no less than its sensing rate, i.e., $R(t) \geq R_s, \forall \gamma(t) = 1$.

## *4.1.2 Problem Formulation*

In this subsection, we first introduce the energy consumption of the UAV, and then formulate the EE maximization problem.

### 4.1.2.1 Energy Consumption

The energy consumption of a UAV contains transmission energy consumption and sensing energy consumption. Let $\delta$ be the length of a time slot. Since the transmission power has been introduced in Sect. 4.1.1.2, the transmission energy consumption of the UAV in time slot $t$ can be written as $E_T(t) = P_T(t) \times \delta$. The sensing energy consumption of the UAV consists of propulsion energy that supports the movement among different tasks, and hovering energy for UAV hovering and data collection. With the help of [12], the propulsion energy of the UAV in time slot $t$ is modeled as

$$E_F(t) = \delta \times \left(\kappa_1 \|\boldsymbol{v}(t)\|^3 + \frac{\kappa_2}{\|\boldsymbol{v}(t)\|} \times \left(1 + \frac{\|\boldsymbol{a}(t)\|^2}{g^2}\right)\right), \tag{4.5}$$

where $\kappa_1$, $\kappa_2$, and $g$ are constants. The hovering energy for the UAV in time slot $t$ can be given as $E_H(t) = P_h \times \delta \times \gamma(t)$, where $P_h$ is the hovering power of the UAV.

#### 4.1.2.2 Problem Description

We denote the time slot at which the UAV finishes all its tasks by $T$, and define the EE of the sensing UAV as the received sensory data at the BS divided by its total energy consumption. The EE of the UAV can be given by

$$EE = \frac{\sum_{t=1}^{T} R(t)}{\sum_{t=1}^{T}(E_F(t) + E_T(t) + E_H(t))}. \tag{4.6}$$

Let $\tau_n$ be the time slot that the UAV starts to perform task $n$. Our objective is to maximize the EE of the UAV by optimizing the transmission power and UAV velocity, which contains the information of trajectory and speed. Thus, the problem can be formulated by

$$\max_{\{P_T(t)\},\{\boldsymbol{v}(t)\}} \quad EE \tag{4.7a}$$

$$\boldsymbol{s.t.}\, z(t) \leq h_{max}, \tag{4.7b}$$

$$\|\boldsymbol{v}(t)\| \leq v_{max}, \tag{4.7c}$$

$$\|\boldsymbol{a}(t)\| \leq a_{max}, \tag{4.7d}$$

$$z(t)\tan(\theta) \geq r, \forall \gamma(t) = 1, \tag{4.7e}$$

$$\sum_{t=\tau_n}^{\tau_n+t_0-1} \|\boldsymbol{v}(t)\| = 0, \forall n \in \mathcal{N}, \tag{4.7f}$$

$$R(t) \geq R_s, \forall \gamma(t) = 1. \tag{4.7g}$$

Constraints on the height, velocity, and accelerated velocity are given in (4.7b), (4.7c), and (4.7d), respectively. Equation (4.7e) is the UAV coverage constraint to perform sensing tasks, and (4.7f) shows that the UAV's velocity is zero when performing tasks. Equation (4.7g) explains that the data transmission can be completed simultaneously with the data collection for each task.

### 4.1.3 *Energy Efficiency Maximization Algorithm*

The optimal solution of problem (4.7) cannot be achieved directly since the objective function (4.7a) is non-convex with respect to $\{P_T(t)\}$. Therefore, we decompose it into two subproblems: UAV sensing optimization for variable $\{\boldsymbol{v}(t)\}$, and UAV transmission optimization for transmission power $\{P_T(t)\}$. In the following, we solve these two subproblems to obtain a sub-optimal solution of problem (4.7).

#### 4.1.3.1 UAV Sensing Optimization

Given the transmission power of each UAV, the maximum EE corresponds to the minimum propulsion energy consumption, and thus, the UAV sensing optimization subproblem can be shown as

$$\min_{\{\boldsymbol{v}(t)\}} \sum_{t=1}^{T} E_F(t), \qquad \boldsymbol{s.t.}\, (4.7\text{b}), (4.7\text{c}), (4.7\text{d}), (4.7\text{e}), (4.7\text{f}). \tag{4.8}$$

Note that the objective function and the constraints only contains the norm of the UAV velocity $\|\boldsymbol{v}(t)\|$. However, we need to optimize both the direction, i.e., UAV trajectory, and the norm, i.e., UAV speed for the velocity variable $\boldsymbol{v}(t)$. Therefore, $\boldsymbol{v}(t)$ cannot be solved directly by solving problem (4.8). In the following, we optimize the trajectory and speed of the UAV separately as below.

1. *Trajectory Optimization*: The minimum propulsion energy consumption can be obtained when the length of the trajectory is minimized. Therefore, we aim to minimize the length of the UAV trajectory for completing these $N$ tasks in this part. We denote the location that the UAV performs the $n$-th sensing task by $\boldsymbol{S}_n$. The trajectory of the UAV between the $n$-th and $(n+1)$-th tasks is given as a line segment $\boldsymbol{S}_{n+1} - \boldsymbol{S}_n$, which is the shortest line between $\boldsymbol{S}_n$ and $\boldsymbol{S}_{n+1}$.

   We define the space where the UAV can perform the $n$-th sensing task as the $n$-th *sensing range*, denoted by $\mathscr{C}_n$. According to the UAV sensing area constraint (4.1), the $n$-th *sensing range* is a conical region, which can be depicted mathematically as

$$(x - x_n)^2 + (y - y_n)^2 \leq ((z - r\cot\theta)\tan\theta)^2. \tag{4.9}$$

   The trajectory optimization problem can be expressed as

$$\min_{\{\boldsymbol{S}_n\}_{n=1}^{N}} \sum_{n=1}^{N} \|\boldsymbol{S}_n - \boldsymbol{S}_{n-1}\|, \tag{4.10a}$$

$$\boldsymbol{s.t.}\, \forall \boldsymbol{S}_n \in \mathscr{C}_n. \tag{4.10b}$$

   Note that the conical region for each *sensing range* is a convex region, and (4.10a) is convex with respect to $\boldsymbol{S}_n$. Therefore, (4.10) is a convex problem, which can be solved by standard convex optimization techniques.
2. *Speed Optimization*: When the moving direction of the UAV is given, we will optimize the speed of the UAV. Without loss of generality, we optimize the speed of the UAV between $\boldsymbol{S}_n$ and $\boldsymbol{S}_{n+1}$ in the following. Since the UAV is static when performing data collection, the initial and final speeds along the designed trajectory should both be 0. Therefore, the speed of the UAV firstly increases

and then decreases along the trajectory. According to the symmetry of (4.5), the energy consumption of the accelerating and decelerating processes is equal. When the speed of the accelerating process is optimized, the optimal UAV speed of the decelerating process can be obtained with negative acceleration. In the following, we only need to optimize the speed in the accelerating process for the first half of the trajectory.

Denote the length of the trajectory obtained by problem (4.10) $\|S_{n+1}-S_n\| = L$, and the distance that the UAV has moved before time slot $t$ as $\lambda(t)$. When a UAV moves along the first half of the trajectory, i.e., $\lambda(t) < L/2$, we optimize the speed of time slot $t$ by a speed recursive function $v(t) = f(v(t-1))$, where $f(\cdot)$ is derived by the following approach. $\lambda(t)$ is then updated as $\lambda(t-1)+v(t)$. When $\lambda(t) \geq L/2$, the UAV starts to decelerate, and the speed is symmetric to the acceleration.

**Theorem 4.1** *The UAV speed in time slot t is set as the value that is closest to $v^{opt}(t)$ within the feasible solution range, where $v^{opt}(t)$ is the positive real root of the following equation:*

$$\kappa_1 g^2(v(t))^4 + \kappa_2 v(t-1)v(t) - \kappa_2(g^2 + (v(t-1))^2) = 0. \tag{4.11}$$

***Proof*** In this part, we derive the UAV speed recursive function $f(\cdot)$. When the most energy efficient speed $v(t)$ is achieved, the UAV moves a unit distance with the minimum energy consumption. Therefore, the energy consumption for moving a distance of $\Delta l$ with $\Delta l \to 0$ is $E_F^\Delta = \frac{\Delta l}{v(t)} \times \frac{E_F(t)}{\delta}$. With the maximum accelerated speed being $a_{max}$, the range of $v(t)$ is given as $v(t-1) - a_{max} \leq v(t) \leq v(t-1) + a_{max}$. Therefore, the problem is converted to finding the minimum $E_F^\Delta$ within a given range of the variable $v(t)$.

The minimum $E_F^\Delta$ is achieved when $\frac{dE_F^\Delta}{dv(t)} = 0$. We then substitute it into $\frac{dE_F^\Delta}{dv(t)} = 0$, and the equation is simplified as given in (4.11). It can be easily known that (4.11) has only one positive real root, denoted by $v^{opt}(t)$ for simplicity, which can be solved by the quartic equation root formula. The energy consumption $E_F^\Delta$ decreases when $0 \leq v(t) \leq v^{opt}(t)$, and increases in the range $v(t) \geq v^{opt}(t)$. □

The algorithm is summarized in Algorithm 3.

#### 4.1.3.2 UAV Transmission Optimization

Given the trajectory and the speed, the transmission power is optimized. Based on (4.3) and (4.4), the UAV transmission power can be expressed as $P_T(t) = \left(2^{R(t)/W_B} - 1\right) \times \sigma^2 \times 10^{PL_a(t)/10}$. We then substitute it into (4.7a), and the UAV transmission optimization subproblem can be converted to

**Algorithm 3:** UAV velocity design algorithm

**begin**
  **Initialization** $\lambda(t) = 0$, $v_i(t) = 0$;
  **while** $\lambda(t) < L/2$ **do**
    $t = t + 1$;
    $v_i(t) = f(v_i(t-1))$;
    $\lambda(t) = \lambda(t-1) + v_i(t)$;
  **end**
  $t_{L/2} = t$;
  **for** $t = (t_{L/2} + 1) : (2 \times t_{L/2})$ **do**
    $v_i(t) = v_i(t_{L/2} - (t - t_{L/2}))$;
  **end**
**end**

$$\max_{\{P_T(t)\}} \frac{\sum_{t=1}^{T} R(t)}{A + B \times \sum_{t=1}^{T} 2^{R(t)/W_B}}, \quad (4.12)$$

$$\boldsymbol{s.t.}\ (4.7\text{g}),$$

where $A = \sum_{t=1}^{T}(E_F(t)+E_H(t)-\sigma^2\times 10^{PL_a(t)/10}\times\delta)$ and $B = \sigma^2\times 10^{PL_a(t)/10}\times \delta$ are constants. In the following, we first find the optimal solution of the objective function, and then give the optimal solution that satisfies constraint (4.7g).

The data rate $R(t)$ with the maximum EE satisfies

$$\partial EE/\partial R(t) = 0, \forall \gamma(t) = 1. \quad (4.13)$$

In addition, the data rate in different time slots is symmetric in (4.12). Therefore, the rates in different time slots are equal when the maximum EE is achieved, i.e.,

$$R(\tau_n) = \ldots = R(\tau_n + t_0 - 1), \forall n \in \mathcal{N}. \quad (4.14)$$

When we substitute (4.12) and (4.14) into (4.13), it can be derived that the optimal rate is the solution of the equation below

$$(\ln 2 \cdot R(t) - 1) \cdot 2^{R(t)/W_B} = A/(B \cdot t_0). \quad (4.15)$$

The left side of Eq. (4.15) increases monotonically, and the right side of Eq. (4.15) is constant. Therefore, Eq. (4.15) has only one root, which can be obtained by numerical methods. For simplicity, we denote the root of Eq. (4.15) by $R_0$.

It is known that the EE increases monotonically when $0 \leq R(t) \leq R_0$, and decreases monotonically when $R_0 \leq R(t)$. To satisfy constraint (4.7g), we have $R(t) \geq R_s, \forall \gamma(t) = 1$. In conclusion, the optimal transmission rate for the UAV is given by $R_n^{opt} = \max\{R_0, R_s\}$. The optimal transmission power can be written as $P_T^{opt}(t) = \left(2^{R_n^{opt}/W_B} - 1\right) \times \sigma^2 \times 10^{PL_a(t)/10}$.

#### 4.1.3.3 Overall Algorithm

The UAV sensing and transmission optimization algorithm can be summarized as below. The UAV first performs trajectory optimization and finds the shortest trajectory that can complete all the tasks. Then, given the trajectory of the UAV, the speed between any of the two sensing tasks can be optimized, as shown in Sect. 4.1.3.1. With given UAV trajectory and speed, the transmission power of the UAV in each time slot can be solved with the UAV transmission optimization algorithm proposed in Sect. 4.1.3.2.

### *4.1.4 Simulation Results*

In this subsection, we evaluate the performance of our proposed algorithm. The selection of the simulation parameters is based on the existing specification [13]. The length of each time slot is considered as $\delta = 1$ s. The initial location of the UAV is randomly distributed in a 3-dimensional space of 2 km $\times$ 2 km $\times$ 200 m. The UAV has 3 tasks to be finished in turn, with $t_0 = 10$. The locations of the tasks are randomly distributed on the ground of the 2 km $\times$ 2 km area, and the projection of the initial location of the UAV to the ground also falls in this area. The hovering power of the UAV is set as $P_h = 50$ W, and the other UAV parameters are given as $a_{max} = 10$ m/s$^2$, $h_{max} = 200$ m, $\theta = 15°$, $\kappa_1 = 9.26 \times 10^{-4}$, $\kappa_2 = 2250$, $g = 9.8$. The channel parameters are given as $\alpha = 12$, $\beta = 0.135$, $\eta_{LoS} = 1$, $\eta_{NLoS} = 20$, $L_{FS} = 32.44$, $W_B = 1$ MHz.

We provide two schemes in comparison with the proposed energy efficiency maximization algorithm (EEMA):

1. **Exhaustive search method (ESM)**: We obtain the maximum EE by enumerating over 1000 trajectory strategies and over 1000 transmission power possibilities for each UAV sensing task.
2. **Uniform accelerate motion method (UAMM)**: The UAV moves along the optimized trajectory with fixed accelerated velocity $a_{max}$, and the UAV transmission optimization is the same with that in EEMA.

Figure 4.2 illustrates the EE with different UAV sensing rates $R_s$, where the maximum UAV speed is set as $v_{max} = 50$ m/s. When $R_s < 12$ Mbps, the EE increases with $R_s$, owing to the decreasing energy consumption proportion for propulsion. The EE decreases with $R_s > 12$ Mbps, which is affected by the exponentially increasing transmission energy consumption. The difference between EEMA and UAMM is smaller with a high sensing rate, which shows that propulsion energy consumption is the major influence factor in the low sensing rate system, while transmission energy consumption is the major influence factor in the high sensing rate system. The average EE performance gap between EEMA and ESM is less than 2% within the simulation range.

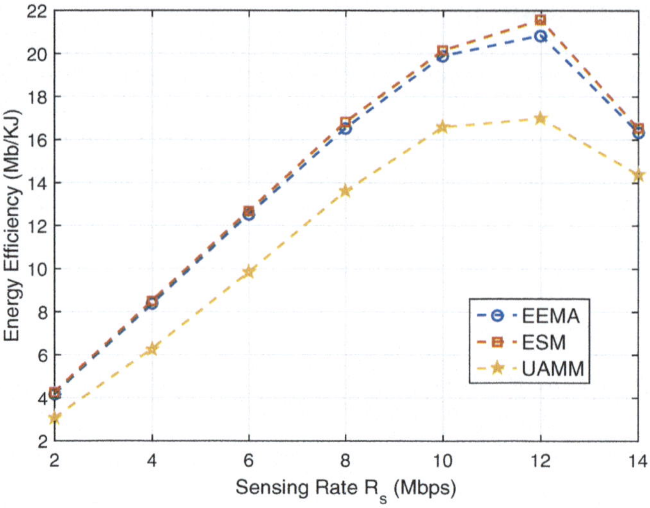


**Fig. 4.2** UAV sensing rate vs. energy efficiency ($v_{max} = 50$ m/s)

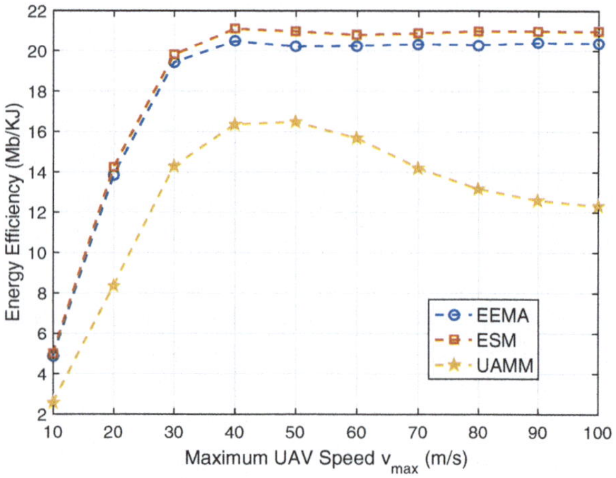


**Fig. 4.3** Maximum UAV speed vs. energy efficiency ($R_s = 10$ Mbps)

Figure 4.3 shows the EE versus the maximum UAV speed $v_{max}$, where the data collection rate $R_s$ is set as 10 Mbps. In the EEMA scheme, the EE increases rapidly with the maximum UAV speed $v_{max}$ when $v_{max} < 40$ m/s. The EE converges to a stable value when the maximum UAV speed $v_{max} > 50$ m/s, which implies that the optimized UAV speed is always no more than 50 m/s. The trend of the ESM scheme is similar with that of the EEMA scheme, and the gap of EE between EEMA and ESM is less than 3%. In the UAMM scheme, the EE first increases when $v_{max} <$ 40 m/s, but is about 20% lower than the EEMA scheme. The EE decreases after $v_{max} > 50$ m/s, as a large UAV speed may have a negative impact on the EE.

### 4.1.5 Summary

In this section, we have studied a cellular UAV system, and proposed the EEMA that contains UAV sensing and transmission optimization to maximize the EE efficiently. Simulation results have shown that the performance gap between the EEMA scheme and ESM is less than 3%. Besides, in the low sensing rate systems, the propulsion energy consumption dominates the EE, while in the high sensing rate systems, the transmission energy consumption dominates the EE.

## 4.2 Cooperative Cellular Internet of UAVs

In the cellular Internet of UAVs, the sensory data is transmitted to the BSs directly through the cellular network [14–16]. Note that sensing failure may occur due to imperfect sensing in practical systems. In this section, we study a cooperative cellular Internet of UAVs to further improve the successful sensing probability. To be specific, multiple UAVs are arranged to collect the sensory data for the same sensing task cooperatively, and to transmit the collected data to the BS separately. In this way, the successful sensing probability requirement for each UAV is loosened [17, 18], and the task completion time of each UAV can be shortened [19]. Such cooperation improves the performance of both UAV sensing and UAV transmission, since the two processes are designed jointly in this network.

Although UAV cooperation has the advantages in reducing the sensing failure probability and task completion time, it also involves some challenges. Firstly, as the UAV scheduling will influence the sensing performance, an efficient scheduling scheme is necessary. Secondly, since each sensing task is performed by multiple UAVs, the trajectories and sensing locations of UAVs are coupled with each other. In light of these issues, we first propose a sense-and-send protocol to support the cooperation and facilitate the scheduling. Then, we optimize the trajectories, sensing locations, and scheduling of these cooperative UAVs to minimize the completion time for all the tasks. As the problem is NP-hard, we decompose it into three subproblems, i.e., trajectory optimization, sensing location optimization, and UAV scheduling, and solve it by an iterative algorithm with a low complexity.

The rest of this section is organized as follows. In Sect. 4.2.1, we describe the system model of the cooperative UAV network. In Sect. 4.2.2, we elaborate on the sense-and-send protocol for the cooperative cellular Internet of UAVs. In Sect. 4.2.3, we formulate the task completion time minimization problem by optimizing the trajectory, sensing location, and UAV scheduling. An ITSSO algorithm is proposed to solve the problem together with the algorithm analysis in Sect. 4.2.4. Simulation results are presented in Sect. 4.2.5, and finally we summarize this section in Sect. 4.2.6.

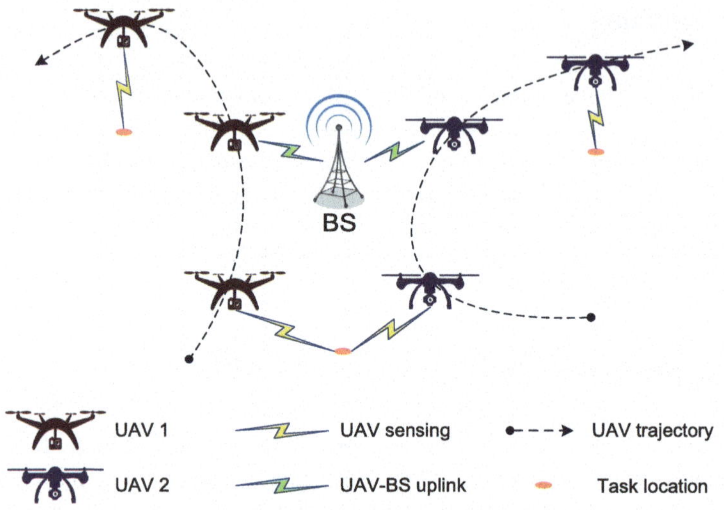


**Fig. 4.4** System model for the cooperative cellular Internet of UAVs

### *4.2.1 System Model*

We consider a single cell OFDM cellular Internet of UAVs network[1] as shown in Fig. 4.4 [20], which consists of one BS, $M$ UAVs, denoted by $\mathscr{M} = \{1, 2, \ldots, M\}$, and $K$ orthogonal subchannels, denoted by $\mathscr{K} = \{1, 2, \ldots, K\}$. Within the cell coverage, there are $N$ sensing tasks to be completed, denoted by $\mathscr{N} = \{1, 2, \ldots N\}$. The UAVs perform each task in two steps: UAV sensing and UAV transmission, and thus, these two procedures will be repeated by the UAVs until all the tasks are completed. Note that different types of sensing tasks require UAVs with different sensors [21, 22]. Therefore, the sensory data of each task is collected by a predefined UAV group cooperatively, and the UAVs in this group send the collected data to the BS separately. We denote the UAV group that performs task $j$ by $\mathscr{W}_j$, satisfying $\mathscr{W}_j \subseteq \mathscr{M}$, and $|\mathscr{W}_j| = q$. For UAV $i$, it is required to execute a subset of tasks in sequence,[2] denoted by $\mathscr{N}_i = \{1, 2, \ldots N_i\}, \forall i \in \mathscr{M}$, with $\mathscr{N}_i \subseteq \mathscr{N}$. In the following, we first describe the UAV sensing and UAV transmission steps, and then introduce the task completion time of the UAVs in the network.

[1]The multiple cell scenario is an extension of the single cell scenario, and will be studied in the future.

[2]An example of tasks in this model is the geological detection. Specifically, each UAV is arranged to perform a series of tasks, and the geological information of each task is sensed by multiple UAVs.

#### 4.2.1.1 UAV Sensing

In UAV sensing, it is important to design the trajectory of each UAV along which it moves towards the locations of a sequence of tasks. Without loss of generality, we denote the location of the BS by $(0, 0, H)$, and the location of task $n$ by $(x^n, y^n, 0)$. In time slot $t$, let $\boldsymbol{l}_i(t) = (x_i(t), y_i(t), z_i(t))$ be the location of UAV $i$, and $\boldsymbol{v}_i(t) = (v_i^x(t), v_i^y(t), v_i^z(t))$ be its velocity, with $\boldsymbol{v}_i(t) = \boldsymbol{l}_i(t) - \boldsymbol{l}_i(t-1)$. Due to the space and mechanical limitations, the UAVs have a maximum velocity $v_{max}$. For safety consideration, we also assume that the altitude of the UAVs in this network should be no less than a minimum threshold $h_{min}$.

In time slot $t$, the distance between UAV $i$ and the BS is expressed as

$$d_{i,BS}(t) = \sqrt{(x_i(t))^2 + (y_i(t))^2 + (z_i(t) - H)^2}. \tag{4.16}$$

The distance between UAV $i$ and task $j$ is given by

$$d_{i,j}(t) = \sqrt{(x_i(t) - x^n)^2 + (y_i(t) - y^n)^2 + (z_i(t))^2}. \tag{4.17}$$

We utilize the probabilistic sensing model as introduced in [23–25], where the successful sensing probability is an exponential function of the distance between the sensing UAV and the task location. The successful sensing probability for UAV $i$ to perform sensing task $j$ can be shown as

$$PR(i, j) = e^{-\lambda d_{i,j}(t)}, \tag{4.18}$$

where $\lambda$ is a parameter evaluating the sensing performance. The probability that task $j$ is successfully sensed can be expressed by

$$PR_j = 1 - \prod_{i \in \mathscr{W}_j} (1 - PR(i, j)). \tag{4.19}$$

Define $PR_{th}$ as the probability threshold, and task $j$ can be considered to be sensed successfully when $PR_j \geq PR_{th}$.

The UAVs collect an amount of sensory data in the UAV sensing process, but the sensing is considered to be successful only when the collected data is valid for a specific task. Note that the validity of data collection cannot be judged by the UAV immediately, and it needs to be verified by the BS with further data processing. Therefore, the sensory data of a UAV needs to be transmitted to the BS to determine whether the UAV sensing is successful or not, and the time duration for each sensing process is pre-settled as a constant.

#### 4.2.1.2 UAV Transmission

In the UAV transmission, the UAVs transmit the sensory data to the BS over orthogonal subchannels to avoid severe interference. We adopt the 3GPP channel model to evaluate the urban macro cellular support for UAVs [13].

Let $P_U$ be the transmit power of each UAV. The received power at the BS from UAV $i$ in time slot $t$ can then be expressed as

$$P_{i,BS}(t) = \frac{P_U}{10^{PL_{a,i}(t)/10}}, \tag{4.20}$$

where $PL_{a,i}(t)$ is the average air-to-ground pathloss, defined by $PL_{a,i}(t) = P_{L,i}(t) \times PL_{L,i}(t) + P_{N,i}(t) \times PL_{N,i}(t)$. Here, $PL_{L,i}(t)$ and $PL_{N,i}(t)$ are the LoS and NLoS pathloss from UAV $i$ to the BS, with $PL_{L,i}(t) = 28+22\times\log(d_{i,BS}(t))+ 20\times\log(f_c)$, and $PL_{N,i}(t) = -17.5+(46-7\times\log(z_i(t)))\times\log(d_{i,BS}(t))+20\times \log(\frac{40\pi\times f_c}{3})$, respectively, where $f_c$ is the carrier frequency. $P_{L,i}(t)$ and $P_{N,i}(t)$ are the probability of LoS and NLoS, respectively, with $P_{N,i}(t) = 1 - P_{L,i}(t)$. The expression of LoS probability is given by

$$P_{L,i}(t) = \begin{cases} 1, & d_i^H(t) \le d_1, \\ \frac{d_1}{d_i^H(t)} + e^{\left(\frac{-d_i^H(t)}{p_0}\right)\left(1-\frac{d_1}{d_i^H(t)}\right)}, & d_i^H(t) > d_1, \end{cases} \tag{4.21}$$

where $p_0 = 4300 \times \log(z_i(t)) - 3800$, $d_1 = \max\{460 \times \log(z_i(t)) - 700, 18\}$, and $d_i^H(t) = \sqrt{(x_i(t))^2 + (y_i(t))^2}$. Note that the cooperative sensing and transmission process is completed in a long time period, and small-scale fading is neglected in the transmission channel model.

Therefore, the SNR from UAV $i$ to the BS is given by

$$\gamma_i(t) = \frac{P_{i,BS}(t)}{\sigma^2}, \tag{4.22}$$

where $\sigma^2$ is the variance of AWGN with zero mean. For fairness consideration, each UAV can be assigned to at most one subchannel. We define a binary UAV scheduling variable $\psi_i(t)$ for UAV $i$ in time slot $t$, where

$$\psi_i(t) = \begin{cases} 1, \text{UAV } i \text{ is paired with a subchannel,} \\ 0, \text{otherwise.} \end{cases} \tag{4.23}$$

Therefore, the data rate from UAV $i$ to the BS is given by

$$R_i(t) = \psi_i(t) \times W_B \log_2(1 + \gamma_i(t)), \tag{4.24}$$

where $W_B$ is the bandwidth of a subchannel.

#### 4.2.1.3 Task Completion Time

For UAV $i$, the relation between two consecutive sensing time slots $\tau_i^j$ and $\tau_i^{j+1}$ is given as

$$\sum_{t=\tau_i^j}^{\tau_i^{j+1}} \boldsymbol{v}_i(t) = \boldsymbol{l}_i(\tau_i^{j+1}) - \boldsymbol{l}_i(\tau_i^j), \forall j \in \mathcal{N}_i, \tag{4.25}$$

where $\boldsymbol{l}_i(\tau_i^j)$ and $\boldsymbol{l}_i(\tau_i^{j+1})$ are the sensing locations of its $j$th and $j+1$th task. We define the task completion time of UAV $i$ as the number of time slots that it costs to complete the sensing and transmission of all its tasks, which can be expressed as

$$T_i = \tau_i^{N_i} + T_{tran,i}^{N_i}, \tag{4.26}$$

where $\tau_i^{N_i}$ is the time slot in which it performs the data collection for its last task, and $T_{tran,i}^{N_i}$ is the time that UAV $i$ costs to complete the data transmission for its last task $N_i$.

### 4.2.2 Sense-and-Send Protocol

In this subsection, we present the sense-and-send protocol for the UAV cooperation. As illustrated in Fig. 4.5, the UAVs perform sensing and data transmission for the tasks in a sequence of time slots. For each UAV, the time slots can be classified

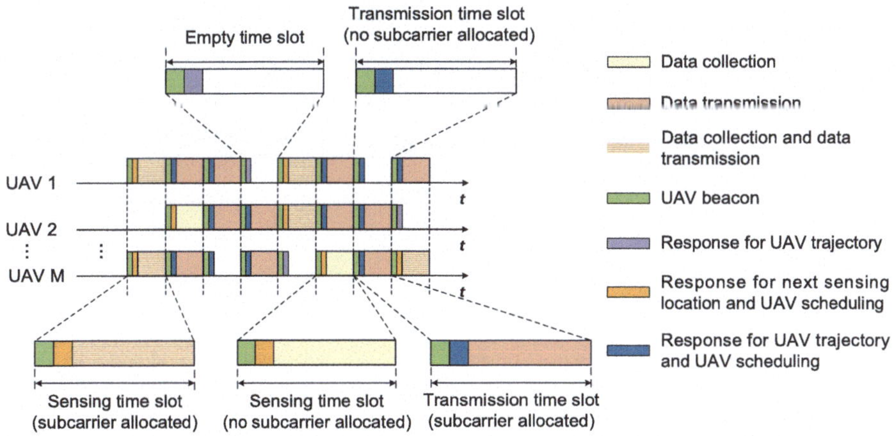


**Fig. 4.5** Sense-and-send protocol

into three types: sensing time slot, transmission time slot, and empty time slot. For convenience, we define the location that the UAV performs a sensing task as the **sensing location**. In each sensing time slot, the UAV collects and transmits data for its current task at the sensing location. In each transmission time slot, the UAV moves along the optimized trajectory and transmits the collected data to the BS. A UAV is in an empty time slot if it has completed data transmission and has not reached the next sensing location. In an empty time slot, the UAV neither collects data nor transmits data, and only moves towards the next sensing location. In the following, we will elaborate on these three types of time slots.

When performing a task, a UAV is in the sensing time slot first, and then switches to the transmission time slot. After several transmission time slots, a UAV may either be in the empty slot or in the next sensing time slot. The BS optimizes the trajectory, sensing location, and UAV scheduling for the UAVs in advance, and responses the required information to the corresponding UAVs in every time slot. The interaction between the BS and UAVs in sensing, transmission, and empty time slots are given in Fig. 4.6, respectively.

**Sensing Time Slot** The UAV hovers on the sensing location and performs data collection and transmission in sensing time slot. As shown in Fig. 4.6a, the UAV first sends UAV beacon to the BS over the control channel, which contains the

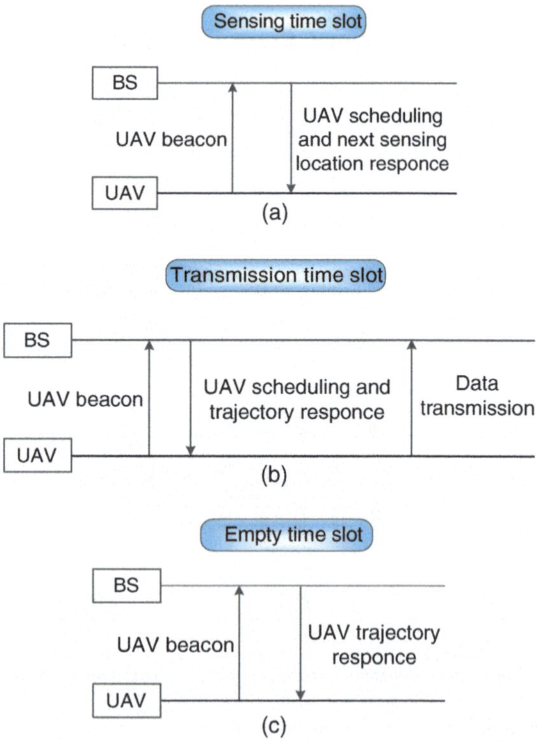


**Fig. 4.6** UAV-BS interaction process in the sense-and-send protocol

information of its location, the ongoing sensing task, the location of the next sensing task, and the transmission request. The BS then informs the UAV of the subchannel allocation result and the sensing location of its next task. Afterwards, the UAV performs data collection until the end of the time slot. The UAV performs data transmission to the BS simultaneously if a subchannel is allocated to it. For task $j$, the UAV hovers on the sensing location to collect data for only one time slot, with a data collection rate $R_s^j$. If the UAV has not finished data transmission in the sensing time slot, it switches to transmission time slot; otherwise, it switches to an empty time slot.

**Transmission Time Slot** When the UAV requires data transmission to the BS, it will operate in the transmission time slot. As shown in Fig. 4.6b, the UAV first sends UAV beacon to the BS over the control channel, which contains the information of transmission request, the UAV location, the data length to transmit, and the location of the next sensing location. The BS then informs the UAV of its trajectory and UAV scheduling solutions in this time slot. Afterwards, each UAV moves along the optimized trajectory. In the meanwhile, the UAV performs data transmission if a subchannel is allocated to it. Otherwise, the UAV cannot transmit data in the current time slot and will send data transmission request again in the next transmission time slot. The collected data with respect to task $j$ should be uploaded to the BS before UAV $i$ starts the next sensing task, i.e., $\sum_{t=\tau_i^j+1}^{\tau_i^{j+1}} R_i(t) \geq R_s^j$, where $\tau_i^j$ is the sensing time slot of UAV $i$ for its $j$th task.

If a UAV does not complete its data transmission in the current time slot, it will occupy another transmission time slot, and request data transmission to the BS again in the next time slot. When a UAV completes data transmission for the current task, it switches to sensing time slot if it has arrived at the sensing location of the next task. Otherwise, it switches to the empty time slot.

**Empty Time Slot** A UAV is in the empty time slot when it has completed the data transmission for the current task, and has not arrived at the next sensing location. As illustrated in Fig. 4.6c, the UAV sends a UAV beacon that contains its current location and its next sensing location to the BS over the control channel. The BS responses the corresponding trajectory to the UAV. The UAV then moves along the optimized trajectory with neither sensing nor transmission in such a time slot. The UAV will switch to the sensing time slot when it arrives at the sensing location of the next task.

*Remark 4.1* In each time slot, at most $K$ UAVs can perform data transmission to the BS. When more than $K$ UAVs send transmission request to the BS in one time slot, the UAVs have to share the subchannels in a time-division multiplexing manner.

To describe the signaling cost over the control channels for the proposed protocol, we assume that the UAV beacon contains no more than $\kappa$ messages, and the trajectory, sensing location, and UAV scheduling response contains at most $\iota$ messages. Therefore, the maximum signaling cost of the network is $M \times (\kappa + \iota)$ in each time slot. The maximum signaling cost is restricted by the number of UAVs, and the signaling of each user costs no more than hundreds of bits [26]. Thus, the signaling cost of the network is tolerable.

## 4.2.3 Problem Formulation

In this part, we first formulate the task completion time minimization problem. Afterwards, we decompose it into three subproblems, and introduce the proposed algorithm that solves the three subproblems iteratively.

### 4.2.3.1 Problem Description

Note that the time for completing all the tasks in this network is determined by the maximum task completion time of the UAVs. Let $T_{max}$ be the maximum task completion time of the UAVs in this network, i.e., $T_{max} = \max\{T_i\}, \forall i \in \mathcal{M}$. To complete all the tasks efficiently, our objective is to minimize the maximum task completion time of the UAVs by optimizing UAV trajectory that consists of speed and direction of the UAV, UAV sensing location, and UAV scheduling. Thus, the problem can be formulated by

$$\min_{\{\boldsymbol{v}_i(t), \boldsymbol{l}_i(\tau_i^j), \psi_i(t)\}} \quad T_{max}, \tag{4.27a}$$

$$\boldsymbol{s.t.}\ z_i(t) \geq h_{min}, \forall i \in \mathcal{M}, \tag{4.27b}$$

$$\|\boldsymbol{v}_i(t)\| \leq v_{max}, \forall i \in \mathcal{M}, \tag{4.27c}$$

$$\|\boldsymbol{v}_i(\tau_i^j)\| = 0, \forall i \in \mathcal{M}, j \in \mathcal{N}_i, \tag{4.27d}$$

$$PR_j \geq PR_{th}, \forall j \in \mathcal{N}, \tag{4.27e}$$

$$\sum_{t=\tau_i^j}^{\tau_i^{j+1}-1} R_i(t) \geq R_s^j, \forall i \in \mathcal{M}, j \in \mathcal{N}_i, \tag{4.27f}$$

$$\sum_{i=1}^{M} \psi_i(t) \leq K, 1 \leq t \leq T_{max}, \tag{4.27g}$$

$$\psi_i(t) = \{0, 1\}. \tag{4.27h}$$

The altitude and velocity constraints are given by (4.27b) and (4.27c), respectively. Constraint (4.27d) shows that the UAV's velocity is zero when performing sensing, and constraint (4.27e) implies that the successful sensing probability for each task should be no less than the given threshold $PR_{th}$. Constraint (4.27f) explains that the data transmission is required to be completed before the next sensing task, and constraint (4.27g) is the UAV scheduling constraint.

#### 4.2.3.2 Problem Decomposition

Problem (4.27) contains both continuous variables $\boldsymbol{v}_i(t)$ and $\boldsymbol{l}_i(\tau_i^j)$, and binary variable $\psi_i(t)$, which is NP-hard. To solve this problem efficiently, we propose an ITSSO algorithm, by solving its three subproblems: trajectory optimization, sensing location optimization, and UAV scheduling iteratively.

In the trajectory optimization subproblem, given the sensing locations $\boldsymbol{l}_i(\tau_i^j), \forall i \in \mathcal{M}, j \in \mathcal{N}_i$ and the UAV scheduling $\psi_i(t), \forall i \in \mathcal{M}$, we can observe that different UAVs are independent, and different tasks of a UAV are irrelevant. Therefore, in this subproblem, the trajectory for a single UAV between two successive tasks can be solved in parallel. Without loss of generality, we study the trajectory for UAV $i$ between its $j$th and $j+1$th task in the rest of this subsection, and the UAV trajectory optimization subproblem can be written as

$$\min_{\boldsymbol{v}_i(t)} (\tau_i^{j+1} - \tau_i^j), \tag{4.28a}$$

$$s.t.\ z_i(t) \geq h_{min}, \forall i \in \mathcal{M}, \tag{4.28b}$$

$$\|\boldsymbol{v}_i(t)\| \leq v_{max}, \tag{4.28c}$$

$$\|\boldsymbol{v}_i(\tau_i^j)\| = 0, \tag{4.28d}$$

$$\sum_{t=\tau_i^j}^{\tau_i^{j+1}-1} R_i(t) \geq R_s^j, \forall i \in \mathcal{M}, j \in \mathcal{N}_i. \tag{4.28e}$$

In the sensing location optimization subproblem, given the UAV scheduling result and trajectory optimization method, the sensing location optimization subproblem can be written as

$$\min_{\{\boldsymbol{l}_i(\tau_i^j)\}} T_{max}, \tag{4.29a}$$

$$s.t.\ \|\boldsymbol{v}_i(t)\| \leq v_{max}, \forall i \in \mathcal{M}, \tag{4.29b}$$

$$\|\boldsymbol{v}_i(\tau_i^j)\| = 0, \forall i \in \mathcal{M}, j \in \mathcal{N}_i, \tag{4.29c}$$

$$PR_j \geq PR_{th}, \forall j \in \mathcal{N}, \tag{4.29d}$$

$$\sum_{t=\tau_i^j}^{\tau_i^{j+1}-1} R_i(t) \geq R_s^j, \forall i \in \mathcal{M}, j \in \mathcal{N}_i. \tag{4.29e}$$

In the UAV scheduling subproblem, given the trajectory optimization and sensing location optimization of each UAV, the UAV scheduling subproblem can be written as

$$\min_{\{\psi_i(t)\}} T_{max}, \tag{4.30a}$$

$$s.t. \sum_{t=\tau_i^j}^{\tau_i^{j+1}-1} R_i(t) \geq R_s^j, \forall i \in \mathcal{M}, j \in \mathcal{N}_i, \tag{4.30b}$$

$$\sum_{i=1}^{M} \psi_i(t) \leq K, 1 \leq t \leq T_{max}, \tag{4.30c}$$

$$\psi_i(t) = \{0, 1\}. \tag{4.30d}$$

#### 4.2.3.3 Iterative Algorithm Description

In this subsubsection, we introduce the proposed ITSSO algorithm to solve problem (4.27), where trajectory optimization, sensing location optimization, and UAV scheduling subproblems are solved iteratively. We first find an initial feasible solution of problem (4.27) that satisfies all its constraints. In the initial solution, the trajectory, sensing location, and UAV scheduling of this solution are denoted by $\{\boldsymbol{v}_i(t)\}^0$, $\{\boldsymbol{l}_i(\tau_i^j)\}^0$, and $\{\psi_i(t)\}^0$, respectively. We set the initial sensing location of task $j$ as $(x^n, y^n, h_{min})$ for all the UAVs in $\mathcal{W}_j$. The initial trajectory for UAV $i$ between tasks $j$ and $j+1$ is set as the line segment between these two sensing locations, with the UAV speed being $v_0$ that satisfies $\frac{v_0}{\boldsymbol{l}_i(\tau_i^{j+1})-\boldsymbol{l}_i(\tau_i^j)} \ll \frac{R_s^j}{R_0}$, where $R_0$ is the average transmission rate of UAV $i$ for task $j$. Initially, the subchannels are randomly allocated to $K$ UAVs in each time slot.

We then perform iterations of trajectory optimization, sensing location optimization, and UAV scheduling until the completion time for all the tasks converges. In each iteration, the trajectory optimization given in Sect. 4.2.4.1 is performed firstly with the sensing location optimization and UAV scheduling results given in the last iteration, and the trajectory variables are updated. Next, the sensing location optimization is performed as shown in Sect. 4.2.4.2, with the UAV scheduling obtained in the last iteration, and the trajectory optimization results. Afterwards, we perform UAV scheduling as described in Sect. 4.2.4.3, given the trajectory optimization and sensing location optimization results. When an iteration is completed, we compare the completion time for all the tasks obtained in the last two iterations. If the completion time for all the tasks does not decrease in the last iteration, the algorithm terminates and the result is obtained. Otherwise, the ITSSO algorithm starts the next iteration.

**Algorithm 4:** Iterative trajectory, sensing, and scheduling optimization algorithm

**Initialization**: Set $r = 0$, find an initial solution of problem (4.27) that satisfies all its constraints, denote the current trajectory, sensing location, and UAV scheduling by $\{\boldsymbol{v}_i(t)\}^0$, $\{\boldsymbol{l}_i(\tau_i^j)\}^0$, and $\{\psi_i(t)\}^0$, respectively;

**while** $T_{max}\left(\{\boldsymbol{v}_i(t)\}^{r-1}, \{\boldsymbol{l}_i(\tau_i^j)\}^{r-1}, \{\psi_i(t)\}^{r-1}\right) - T_{max}\left(\{\boldsymbol{v}_i(t)\}^{r}, \{\boldsymbol{l}_i(\tau_i^j)\}^{r}, \{\psi_i(t)\}^{r}\right) > 0$ **do**

- $r = r + 1$;
- Solve the trajectory optimization subproblem, given $\{\boldsymbol{l}_i(\tau_i^j)\}^{r-1}$ and $\{\psi_i(t)\}^{r-1}$;
- Solve the sensing location optimization subproblem, given $\{\boldsymbol{v}_i(t)\}^{r}$ and $\{\psi_i(t)\}^{r-1}$;
- Solve the UAV scheduling subproblem, given $\{\boldsymbol{v}_i(t)\}^{r}$ and $\{\boldsymbol{l}_i(\tau_i^j)\}^{r}$;

**end**

**Output**:$\{\boldsymbol{v}_i(t)\}^{r}, \{\boldsymbol{l}_i(\tau_i^j)\}^{r}, \{\psi_i(t)\}^{r}$;

$T_{max}\left(\{\boldsymbol{v}_i(t)\}^{r}, \{\boldsymbol{l}_i(\tau_i^j)\}^{r}, \{\psi_i(t)\}^{r}\right)$ is defined as the optimization objective function after the $r$th iteration. In iteration $r$, the trajectory optimization variables $\{\boldsymbol{v}_i(t)\}$, the sensing location optimization variables $\{\boldsymbol{l}_i(\tau_i^j)\}$, and the UAV scheduling variables $\{\psi_i(t)\}$ are denoted by $\{\boldsymbol{v}_i(t)\}^r$, $\{\boldsymbol{l}_i(\tau_i^j)\}^r$, and $\{\psi_i(t)\}^r$, respectively. The ITSSO algorithm is summarized in detail as shown in Algorithm 4.

## 4.2.4 *Iterative Trajectory, Sensing, and Scheduling Optimization Algorithm*

In this subsection, we first introduce the algorithms that solve the three subproblems (4.28), (4.29), and (4.30), respectively. Afterwards, we analyze the performance of the proposed ITSSO algorithm.

### 4.2.4.1 Trajectory Optimization

In this subsubsection, we provide a detailed description on the UAV trajectory optimization algorithm (4.28). Note that we utilize the standard aerial vehicular channel fading model as proposed in [13], which makes constraint (4.28e) very complicated. Therefore, problem (4.28) cannot be solved with the existing optimization methods. In the following, we optimize the speed and moving direction of the UAVs with a novel algorithm utilizing geometry theorems and extremum principles.

1. *UAV Speed Optimization*: Assume that the transmission distance from a UAV to the BS is much larger than the UAV velocity, i.e., $d_{i,BS}(t) \gg v_{max}$, $\forall i \in \mathcal{M}$, we have the following theorem on the UAV speed.

**Theorem 4.2** *The optimal solution can be achieved when the speed of the UAV is* $v_{max}$ *between* $\tau_i^j$ *and* $\tau_i^{j+1}$, *i.e.,* $\|\boldsymbol{v}_i(t)\| = v_{max}, \tau_i^j \leq t \leq \tau_i^{j+1}, \forall i \in \mathscr{M}, j \in \mathscr{N}_i$.

***Proof*** We assume that in the optimal trajectory, there exists a time slot $t_0$, in which the speed of the UAV is $\|\boldsymbol{v}_i(t_0)\| = v'$, with $v' < v_{max}$. In the following, we will prove that there exists a solution with $\|\boldsymbol{v}_i(t_0)\| = v_{max}$, whose performance is no worse than the one with $\|\boldsymbol{v}_i(t_0)\| = v'$.

Given $\boldsymbol{l}_i(t_0 - 1)$ and $\boldsymbol{l}_i(t_0 + 1)$, the possible location of $\boldsymbol{l}_i(t_0)$ is shown as the two red points in Fig. 4.7a. When we set $\|\boldsymbol{v}_i(t_0)\| = v_{max}$, the possible location of the UAV in time slot $t_0$ moves to the blue points. If the BS is on the right side of the polyline, at least one blue point is nearer to the BS than both of the red points. Therefore, setting $\|\boldsymbol{v}_i(t_0)\| = v_{max}$ can improve the transmission rate in time slot $t_0$ when the BS is on the right side of the polyline.

Similarly, we can analyze the possible location of $\boldsymbol{l}_i(t_0 - 1)$ in Fig. 4.7b. Given the location of $\boldsymbol{l}_i(t_0 - 2)$ and $\boldsymbol{l}_i(t_0)$, the possible location of $\boldsymbol{l}_i(t_0 - 1)$ is shown as the two red points with $\|\boldsymbol{v}_i(t_0)\| = v'$. When we set $\|\boldsymbol{v}_i(t_0)\| = v_{max}$, the possible location of the UAV in time slot $t_0 - 1$ moves to the blue points. Similarly, we can get a polyline, on the left side of which at least one blue point is nearer to the BS than both of the red points. Thus, the transmission rate in time slot $t_0 - 1$ can be improved when the BS is on the left side of the new polyline.

It can be easily proved that the two polylines are not parallel, and a quadrangle is obtained with the two polylines. Since the transmission distance is much larger than the UAV velocity, i.e., $d_{i,BS}(t) \gg v_{max}$, the BS is outside of the quadrangle. Therefore, the transmission rate in at least one time slot can be improved when $\|\boldsymbol{v}_i(t)\|$ is changed from $v'$ to $v_{max}$. A larger transmission rate reduces the number of transmission time slots, and thus, the task completion time is no more than the solution with $\|\boldsymbol{v}_i(t_0)\| = v'$. □

According to the proof of Theorem 4.2, we have the following remark.

*Remark 4.2* With $t'$ being the optimal solution, a trajectory with the length of $t' \times v_{max}$ can be given.

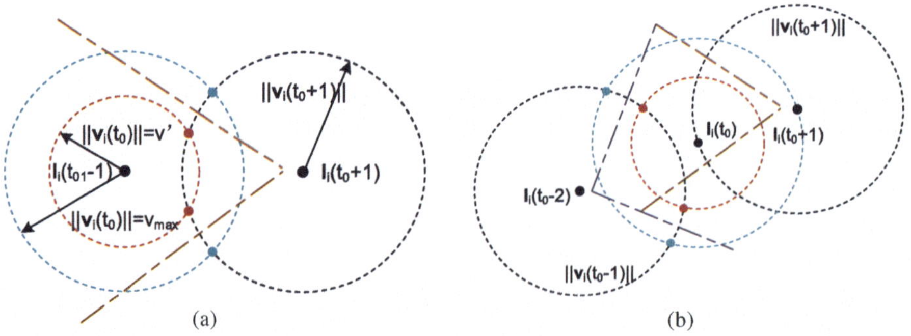


**Fig. 4.7** Proof of Theorem 4.2

Therefore, we set the UAV speed as $v_{max}$ in the following parts.

2. *UAV Flying Direction Optimization*: Since the speed of the UAV has been obtained by Theorem 4.2, we then propose an efficient method to solve the flying direction of UAV $i$ between $\tau_i^j$ and $\tau_i^{j+1}$.

For simplicity, we denote the time between UAV $i$'s $j$th and $j+1$th task by $\delta_i^j = \tau_i^{j+1} - \tau_i^j - 1$. Let $\lceil x \rceil$ be the minimum integer that is no smaller than $x$. In problem (4.28), the lower bound of $\delta_i^j$ can be expressed as

$$\delta_i^{j,lb} = \frac{\boldsymbol{l}_i(\tau_i^{j+1}) - \boldsymbol{l}_i(\tau_i^j)}{v_{max}}, \tag{4.31}$$

which corresponds to a line segment trajectory, with $\boldsymbol{l}_i(\tau_i^{j+1}) - \boldsymbol{l}_i(\tau_i^j)$ being its flying direction. This direction is the solution if constraint (4.28e) can be satisfied, i.e., $\sum_{t=\tau_i^j}^{\tau_i^j+\delta_i^{j,lb}} R_i(t) \geq R_s^j$. Otherwise, the UAV has to make a detour to approach the BS for a larger transmission rate, which also leads to a larger task completion time.

As illustrated in Fig. 4.8, the flying direction of the detoured trajectory contains two parts, namely sending-priority detour and sensing-priority route. In the sending-priority detour part, the UAV detours to the BS for a larger transmission rate, and in the sensing-priority route part, the UAV moves toward its next sensing location with the shortest time.

- **Sending-Priority Detour**: In the sending-priority detour part, we maximize the transmission rate of the UAV, so that constraint (4.28e) can be satisfied with the minimum time slots. To achieve the maximum achievable rate, the UAV moves along the direction with the maximum rate ascent velocity, i.e., the gradient of the transmission rate $R_i(t)$, which can be expressed as

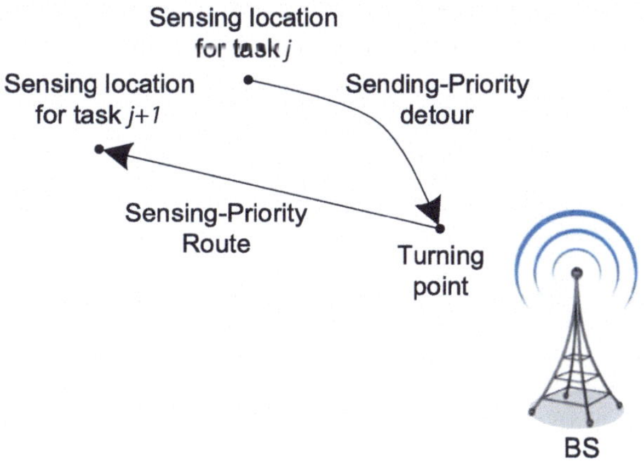


**Fig. 4.8** Flying direction optimization

$$\nabla R_i(t) = (\frac{\partial R_i(t)}{\partial x}, \frac{\partial R_i(t)}{\partial y}, \frac{\partial R_i(t)}{\partial z}), (\boldsymbol{l}_i(t) > h_{min}), \tag{4.32}$$

where the expression of $R_i(t)$ can be derived by Eqs. (4.16), (4.17), (4.20)–(4.24). In time slot $t$, if the altitude of the trajectory is below the minimum threshold $h_{min}$, the UAV has to adjust its flying direction to

$$\nabla R_i(t) = (\frac{\partial R_i(t)}{\partial x}, \frac{\partial R_i(t)}{\partial y}, 0). \tag{4.33}$$

- **Sensing-Priority Route**: We define the endpoint of the sending-priority detour as the *turning point*, denoted by $\boldsymbol{l}_i^{tr}(\tau_i^j)$. In the sensing-priority route part, the UAV flies from the turning point to the sensing location of the next task. To minimize the task completion time, the trajectory of the this part is optimized as a line segment, with $\boldsymbol{l}_i(\tau_i^{j+1}) - \boldsymbol{l}_i^{tr}(\tau_i^j)$ being its flying direction.

We denote the time duration of the sending-priority detour and the sensing-priority route by $\delta_i^{j,1}$ and $\delta_i^{j,2}$, respectively. Our target is to find the minimum $\delta_i^{j,1} + \delta_i^{j,2}$ that satisfies constraint (4.28e). We can observe that a larger $\delta_i^{j,2}$ implies a larger $\delta_i^{j,1}$ since $\delta_i^{j,1}$ is positively related with the detour distance to the BS. Therefore, the minimum $\delta_i^{j,1} + \delta_i^{j,2}$ is achieved with the minimum feasible $\delta_i^{j,1}$. The solution of the flying direction optimization problem is summarized in Algorithm 5.

#### 4.2.4.2 Sensing Location Optimization

In this subsection, we propose a method to solve the sensing location optimization subproblem (4.29). It is known that constraint (4.29e) is non-convex, and thus problem (4.29) cannot be solved directly. Since the trajectory between two consecutive sensing locations has been optimized in Sect. 4.2.4.1, the UAV trajectory is one-to-one correspondence with the sensing locations. Note that constraints (4.29b), (4.29c), and (4.29e) can be satisfied with the trajectory optimization method proposed in Sect. 4.2.4.1 when a sensing location optimization method is given. In

**Algorithm 5:** Flying direction optimization

```
1 Initialization: Set flying direction as l_i(τ_i^{j+1}) − l_i(τ_i^j), and δ_i^{j,1} = 0;
2 while Constraint (4.28e) is not satisfied do
3 |   δ_i^{j,1} = δ_i^{j,1} + 1;
4 |   Optimize the flying direction of sending-priority detour;
5 |   Find the location of the turning point;
6 |   Optimize the flying direction of sensing-priority route;
7 end
8 Set the current flying direction as the final solution;
```

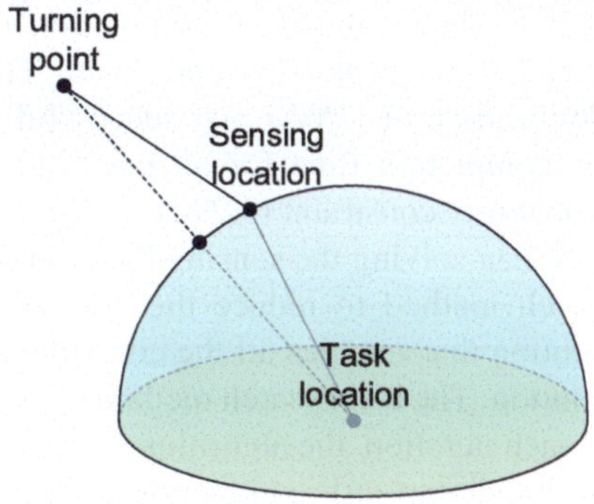


**Fig. 4.9** Illustration of Theorem 4.3

the following, we first analyze the properties of the sensing location, and then solve this subproblem with a local search method.

**Theorem 4.3** *For each UAV, the optimal sensing location is collinear with the corresponding turning point and the task location.*[3]

***Proof*** To satisfy constraint (4.29d), we assume that the distance between the sensing location a UAV and the task location should be no more than $d'$ while considering the sensing locations of other UAVs are fixed. As shown in Fig. 4.9, the feasible solutions of the sensing location is a hemispheroid, with the task location being the center and $d'$ being the radius. We can observe that the sensing location with the shortest trajectory is on the intersection of the hemispheroid and the line segment from the turning point to the task location. Therefore, when the task completion time of a UAV is minimized, the sensing location is collinear with the turning point and the task location. □

In the following, we derive the upper bound and lower bound of $\delta_i^j$, $\forall i \in \mathcal{M}$, $j \in \mathcal{N}_i$.

**Theorem 4.4** *The lower bound of $\delta_i^j$, denoted by $\delta_i^{j,lb}$, is achieved when the sensing location and the turning point are overlapped. The upper bound of $\delta_i^j$, denoted by $\delta_i^{j,ub}$, is achieved when the sensing location of task $j+1$ is $(x^{j+1}, y^{j+1}, h_{min})$.*

***Proof*** When the sensing location and the turning point are overlapped, the UAV trajectory is the gradient of the transmission rate, which corresponds to the maximum achievable transmission rate. Therefore, constraint (4.29e) can be satisfied with minimum number of transmission time slots. When the sensing location of task $j+1$ is $(x^{j+1}, y^{j+1}, h_{min})$, the distance between sensing location and task location is minimized, and the successful sensing probability is maximized. On this condition, the length of the trajectory is maximized, which corresponds to the maximum $\delta_i^j$. □

[3]If the UAV trajectory does not detour to the BS, the start point can be considered as the turning point.

It is shown that task completion time $T_i$ and successful sensing probability $PR(i, j)$ are negatively correlated. Therefore, a trade-off between the task completion time of UAV $i$ and successful sensing probability is required to minimize the completion time for all the tasks while guaranteeing the successful sensing probability constraint (4.29d).

When solving the sensing location optimization subproblem, we propose a local search method to reduce the computational complexity. We first give an initial solution that satisfies all the constraints of problem (4.29), and set it as the current solution. The local search method contains iterations of sensing location adjustment. In each iteration, the algorithm is performed task by task. For task $j$, we first reduce the maximum task completion time of the UAVs in set $\mathscr{W}_j$ by one time slot, and then adjust the sensing location of other UAVs in set $\mathscr{W}_j$ to keep constraint (4.29d) satisfied. The local search method terminates when the completion time for all the tasks cannot be reduced.

The detailed process of the sensing location optimization method is summarized in Algorithm 6. When performing the sensing location adjustment for task $j$, without loss of generality, we assume that the maximum task completion time of the UAVs in $\mathscr{W}_j$ is $T_i$. If $\delta_i^j > \delta_i^{j,lb}$, we reduce $\delta_i^j$ by one time slot by adjusting its sensing location. Denote $\mathscr{U}$ by the set of UAVs in $\mathscr{W}_j$ that satisfies $\delta_m^j \leq \delta_m^{j,ub}$, and $T_m \leq T_i - 1$, so that the maximum task completion time of the UAVs in $\mathscr{W}_j$ does not increase. We adjust the sensing locations of the UAVs in set $\mathscr{U}$ to decrease the distances between their sensing locations and the task location until constraint (4.29d) is satisfied. Denote the task completion time increment of UAV $m$ by $\Delta t_m$, its sensing location after adjustment is given as $\boldsymbol{l}_i(\tau_m^j) = \boldsymbol{l}_i(\tau_m^j) + \Delta t_m \cdot v_{max} \cdot \frac{\boldsymbol{l}_m(\tau_m^j) - \boldsymbol{l}_m^{tr}(\tau_m^{j-1})}{|\boldsymbol{l}_m(\tau_m^j) - \boldsymbol{l}_m^{tr}(\tau_m^{j-1})|}$, $\forall m \in \mathscr{W}_j$. If constraint (4.29d) cannot be satisfied until $\mathscr{U}$ is empty, the maximum task completion time $T_i$ cannot be reduced by adjusting the sensing locations of UAVs in $\mathscr{W}_j$.

**Theorem 4.5** *The iterative sensing location optimization method is convergent.*

***Proof*** For task $j$, the maximum task completion time of the UAVs in $\mathscr{W}_j$ may reduce or remain unchanged in each iteration. Therefore, the time to complete all the tasks does not increase with the iterations. It is known that the time to complete all the tasks has a lower bound. Therefore, the maximum task completion time cannot reduce infinitely, and the iterative sensing location optimization method is convergent. □

**Theorem 4.6** *The complexity of the iterative sensing location optimization method is $O(NqMN_i)$.*

***Proof*** In each iteration of the sensing location optimization method, each of the $N$ tasks is visited for one time, and the total number of UAV sensing for the $N$ tasks is $Nq$. Therefore, the complexity of each iteration is $Nq$. The number of iterations is determined by the number of task completion time reduction when performing the local search algorithm. It is known that each UAV has a lower bound of its

---

**Algorithm 6:** Sensing location optimization

---

**Initialization:** Give a set of initial sensing locations $\{\boldsymbol{l}_i(\tau_i^j)\}$;
**while** *The maximum task completion time of any task is reduced in the last iteration* **do**
  **for** $j = 1 : N$ **do**
    Find the maximum task completion time $T_i, \forall i \in \mathscr{W}_j$;
    **if** $\delta_i^j > \delta_i^{j,lb}$ **then**
      $\delta_i^j = \delta_i^j - 1$;
      Adjust UAV $i$'s sensing location $\boldsymbol{l}_i(\tau_i^j)$;
      **if** *Constraint (4.29d) is satisfied* **then**
        Continue;
      **else**
        **while** *Constraint (4.29d) is not satisfied* **do**
          **if** $\mathscr{U} = \varnothing$ **then**
            Break;
          **end**
          **for** $m \in \mathscr{U}$ **do**
            $\delta_m^j = \delta_m^j + 1$;
            $\boldsymbol{l}_i(\tau_m^j) = \boldsymbol{l}_i(\tau_m^j) + v_{max} \cdot \frac{\boldsymbol{l}_m(\tau_m^j) - \boldsymbol{l}_i^{tr}(\tau_m^{j-1})}{|\boldsymbol{l}_m(\tau_m^j) - \boldsymbol{l}_m^{tr}(\tau_m^{j-1})|}$;
          **end**
        **end**
      **end**
      **if** *Constraint (4.29d) cannot be satisfied* **then**
        Recover the adjusted data and continue;
      **end**
    **end**
  **end**
**end**

---

task completion time, and the task completion time reduction of a UAV is in direct proportion to its task number $N_i$. Therefore, the total number of task completion time reduction is no more than $MN_i$, and the number of iteration is no more than $MN_i$. In conclusion, the complexity of the iterative sensing location optimization method is $O(N\eta MN_i)$. □

#### 4.2.4.3 UAV Scheduling

In this subsubsection, we introduce the UAV scheduling method that solves subproblem (4.30). Problem (4.30) is NP-hard since $\psi_i(t)$ is a discrete variable. In the following, we propose an efficient method that performs UAV scheduling time slot by time slot.

In each time slot, if the BS receives no more than $K$ transmission requests, each of the request UAV will be allocated with one subchannel. In the time slot that the BS receives more than $K$ transmission requests, the BS allocates the subchannels to these $K$ requesting UAVs with maximum task completion time. The task completion

**Algorithm 7:** UAV scheduling algorithm

```
for t = 1 : T_max do
    if Transmission request number < K then
        Allocate a subchannel to each of the request UAV;
    else
        Allocate a subchannel to the K requesting UAVs with maximum task completion
        time;
    end
    Update the task completion time of each UAV with the new UAV scheduling;
end
```

time of each UAV is then updated with the change of UAV scheduling in this time slot, and then the BS continues the UAV scheduling of the next time slot. The process of UAV scheduling is shown in Algorithm 7.

#### 4.2.4.4 Performance Analysis

In this subsection, we first analyze the performance of the proposed ITSSO algorithm, including its convergence and complexity, and then analyze the system performance of the network.

1. *Convergence*:

**Theorem 4.7** *The proposed ITSSO algorithm is convergent.*

***Proof*** As shown in Sect. 4.2.4.2, given the trajectory optimization method, the time for completing all the tasks does not increase with the sensing location optimization. In the UAV scheduling, the time to complete all the tasks decreases after the BS rearranges the subchannels. Therefore, the time for completing all the tasks does not increase with the iterations of the ITSSO algorithm. It is known that the time for completing all the tasks has a lower bound in such a network, and the objective function cannot decrease infinitely. The time for completing all the tasks will converge to a stable value after limited iterations, i.e., the proposed ITSSO algorithm is convergent. □

2. *Complexity*:

**Theorem 4.8** *The complexity of the proposed ITSSO algorithm is* $O(N^2 qMN_i)$.

***Proof*** The proposed ITSSO algorithm consists of iterations of trajectory optimization, sensing location optimization, and UAV scheduling. In each iteration, the complexity of trajectory optimization is $O(M)$, and the complexity of UAV scheduling is $O(KT_{max}) = O(KN_i)$. The complexity of sensing location optimization is $O(NqMN_i)$, which is proved in Theorem 4.6. The number of ITSSO algorithm iterations is relevant to the reduction of the time for completing all the tasks, which is in direct proportion to the number of tasks $N$. Therefore, the complexity of the proposed ITSSO algorithm is $O(N(M + KN_i + NqMN_i)) = O(N^2qMN_i)$. □

3. *System Performance Analysis*: In this part, we analyze the impact of the number of cooperative UAVs $q$ and the sensing probability threshold $PR_{th}$ on the task completion time of the UAVs in the network.

**Theorem 4.9** *The average rate of change of the task completion time $T_{max}$ to the number of cooperative UAVs $q$ can be expressed by*

$$\frac{\Delta T_{max}}{\Delta q} = \frac{(1 - PR_{th})^{1/q} \ln(1 - PR_{th})}{\lambda(1 - (1 - PR_{th})^{1/q})q^2} \times \frac{N_i}{v_{max}}. \tag{4.34}$$

***Proof*** According to Eqs. (4.18) and (4.19), a larger number of cooperative UAVs $q$ requires a lower successful sensing probability for each UAV. Given the sensing probability threshold $PR_{th}$, when a UAV performs data collection, the distance between the UAV and the sensing task can be longer with a larger $q$. Considering the average rate of change, we assume that the distance between every UAV and the sensing task is the same when performing data collection, and the sensing probability of each task equals the sensing probability threshold $PR_{th}$, i.e.,

$$d_{i,j}(\tau_i^j) = d_0, \forall i, m \in \mathscr{W}_j, \tag{4.35}$$

$$PR_j = PR_{th}, \forall j \in \mathscr{N}. \tag{4.36}$$

When substituting (4.35) and (4.36) into (4.18) and (4.19), we have

$$PR_{th} = 1 - (1 - e^{-\lambda d_0})^q. \tag{4.37}$$

The average rate of change of the sensing distance to the number of cooperative UAVs can be achieved by the derivation of $q$ in Eq. (4.37), which is shown as

$$\frac{\Delta d_0}{\Delta q} = -\frac{(1 - PR_{th})^{1/q} \ln(1 - PR_{th})}{\lambda(1 - (1 - PR_{th})^{1/q})q^2}. \tag{4.38}$$

For each UAV, the increment of sensing distance is equal to the decrement of moving distance. Given the UAV speed $v_{max}$, the average change on the completion time for each task with the number of cooperative UAVs is $\frac{\Delta \delta_i^j}{\Delta q} = -\frac{\Delta d_0}{v_{max} \Delta q}$. For UAV $i$, the average rate of change of its task completion time to the number of cooperative UAVs is $\frac{\Delta T_i}{\Delta q} = -\frac{N_i \Delta d_0}{v_{max} \Delta q} = \frac{(1-PR_{th})^{1/q} \ln(1-PR_{th})}{\lambda(1-(1-PR_{th})^{1/q})q^2} \times \frac{N_i}{v_{max}}$. □

Since the number of cooperative UAVs $q$ is positively related with the number of UAVs $M$, we can conclude that the effect of number of UAVs $M$ on the completion time according to Theorem 4.9, which is shown as the following remark.

*Remark 4.3* The task completion time is negatively related with the number of UAVs $M$, and the marginal decrement reduces with the number of UAVs.

**Theorem 4.10** *The average rate of change of UAV i's task completion time $T_i$ to the sensing probability threshold $PR_{th}$ is*

$$\frac{\Delta T_{max}}{\Delta PR_{th}} = \frac{(1 - PR_{th})^{1/q-1}}{\lambda q(1 - (1 - PR_{th})^{1/q})} \times \frac{N_i}{v_{max}}. \tag{4.39}$$

***Proof*** Similar with the proof of Theorem 4.9, we assume that the distance between every UAV and the sensing task is the same when performing data collection, and the sensing probability of each task equals the sensing probability threshold $PR_{th}$. Therefore, Eqs. (4.35), (4.36), and (4.37) are also satisfied. The average rate of change of the sensing distance for one task to the sensing probability threshold can be achieved by the derivation of $PR_{th}$ in Eq. (4.37), which is shown as

$$\frac{\Delta d_0}{\Delta PR_{th}} = -\frac{(1 - PR_{th})^{1/q-1}}{\lambda q(1 - (1 - PR_{th})^{1/q})}. \tag{4.40}$$

For each UAV, the increment of sensing distance is equal to the decrement of moving distance. Given UAV speed $v_{max}$, the average rate of change of the time for each task to the sensing probability threshold is $\frac{\Delta \delta_i^j}{\Delta PR_{th}} = -\frac{\Delta d_0}{v_{max}\Delta PR_{th}}$. For UAV $i$, the average rate of change of its task completion time to the sensing probability threshold is $\frac{\Delta T_i}{\Delta PR_{th}} = -\frac{N_i \Delta d_0}{v_{max}\Delta PR_{th}} = \frac{(1-PR_{th})^{1/q-1}}{\lambda q(1-(1-PR_{th})^{1/q})} \times \frac{N_i}{v_{max}}$. □

In the following, we analyze the dominated factor on the completion time for all the tasks.

**Theorem 4.11** *The transmission resource is a dominated factor on the completion time for all the tasks when the network is crowded.*

***Proof*** We denote the possibility that a UAV in transmission time slot is allocated with a subchannel by $p_t$, with $p_t \propto K/M$. Therefore, the average time that a UAV finishes data transmission for a task is $\eta \times \frac{M}{K}$, where $\eta$ is a proportionality coefficient. The average time that a UAV costs to finish a task can be given as $\max\{\eta \times \frac{M}{K}, \bar{\delta}^{lb}\}$, where $\bar{\delta}^{lb}$ is the average lower bound of the time that a UAV costs to finish a task. Therefore, the transmission resource is the dominated factor on the completion time for all the tasks when $K$ satisfies $\eta \times \frac{M}{K} > \bar{\delta}^{lb}$, i.e., $K < \frac{\eta \times M}{\bar{\delta}^{lb}}$. □

We then discuss the impact of sensing task size $R_s^j$ on the completion time for all the tasks in different transmission resource schemes.

1. **High Transmission Resource**: In high transmission resource scheme, most of the UAVs in transmission time slot are allocated with a subchannel. The impact of sensing task size $R_s^j$ on the completion time for all the tasks is not significant when $R_s^j$ is at a low level, since most of the data transmission tasks can be

**Table 4.1** Dominated factor on the completion time for all the tasks

| | | Transmission resource $K$ | |
|---|---|---|---|
| | | High | Low |
| Sensing data size $R_s^j$ | High | $R_s^j$ | $K$ |
| | Low | Neither | $R_s^j$ & $K$ |

completed without a detour trajectory. When $R_s^j$ is at a high level, the UAVs are more likely to detour to the BS for data transmission, and $R_s^j$ becomes a dominated factor on the completion time for all the tasks.

2. **Low Transmission Resource**: In low transmission resource scheme, the sub-channels are occupied by the UAVs in most of the time slots. On this condition, the task completion time of the UAVs is extended, and most of the UAVs detour to the BS for data transmission. A larger sensing task size $R_s^j$ corresponds to a longer sensing-priority detour to the BS, i.e., $\frac{\partial T_{max}}{\partial R_s^j} > 0$. With the increment of sensing-priority detour, the UAV moves closer to the BS, which improves the average data rate. Therefore, we have $\frac{\partial^2 T_{max}}{\partial (R_s^j)^2} < 0$, and the sensing task size $R_s^j$ has a more significant impact on the completion time for all the tasks $T_{max}$ when it is at a low level.

The dominated factor on the completion time for all the tasks is summarized in Table 4.1. The transmission resource $K$ is a dominated factor on the completion time for all the tasks when it is at a low level. The sensing data size $R_s^j$ is a dominated factor of the completion time for all the tasks when $R_s^j$ and $K$ are both at a high level or low level.

### 4.2.5 Simulation Results

In this section, we evaluate the performance of the proposed ITSSO algorithm. The selection of the simulation parameters is based on the existing 3GPP specifications [13] and works [16]. For comparison, the following schemes are also performed:

- **Non-Cooperative (NC) Scheme**: In NC scheme, each task is required to be completed with only one UAV, i.e., $q = 1$, and the number of tasks in the network is the same with the proposed ITSSO scheme. The proposed trajectory optimization, sensing location optimization, and UAV scheduling methods are also performed in NC scheme.
- **Fixed Sensing Location (FSL) Scheme**: The FSL is given as mentioned in [27]. In FSL scheme, the sensing locations of the UAVs are given as the location right over the locations of the corresponding tasks, with fixed height $H_{FSL} = 50$ m, and the sensing probability constraint (4.27e) is not considered in this scheme. The task arrangement for each UAV in FSL scheme is the same with the proposed ITSSO scheme, and the proposed trajectory optimization and UAV scheduling methods are utilized in FSL scheme.

**Table 4.2** Simulation parameters

| Parameter | Value |
|---|---|
| BS height $H$ | 25 m |
| Carrier frequency $f_c$ | 2 GHz |
| Number of subchannels $K$ | 10 |
| Bandwidth of each subchannel $W_B$ | 1 MHz |
| Sensing task size $R_s$ | 20 Mbps |
| Noise variance $\sigma^2$ | −96 dBm |
| UAV transmit power $P_U$ | 23 dBm |
| Maximum UAV velocity $v_{max}$ | 50 m/s |
| Minimum UAV altitude $h_{min}$ | 10 m |
| Sensing performance parameter $\lambda$ | 0.01 |
| Sensing probability threshold $PR_{th}$ | 0.9 |

For simulation setup, the initial location of the UAVs is randomly and uniformly distributed in a 3-dimensional area of 500 m × 500 m ×100 m, and the tasks are uniformly distributed on the ground of this area. We assume that the number of tasks for different UAVs is equal, i.e., $N_i = N_j = \frac{N \times q}{M}, \forall i, j \in \mathscr{M}$, and the task arrangement for each UAV is given randomly. The data collection rate for different tasks is fixed, denoted by $R_s^j = R_s, \forall j \in \mathscr{N}$. The values of $N$, $q$, and $m$ are given in each figure. All curves are generated by over 10,000 instances of the proposed algorithm. The simulation parameters are listed in Table 4.2.

Figure 4.10 depicts the completion time for all the tasks $T_{max}$ v.s. the number of cooperative UAVs $q$ for each task. The number of tasks in the network is set as 20, and each UAV is arranged to complete 4 tasks. The number of UAVs $M$ varies with $q$, satisfying $M \times N_i = N \times q$. It is shown that the completion time for all the tasks decreases with the increment of cooperative UAVs for a task. The reason is that the average distance between the UAV and the task increases with a larger number of cooperative UAVs $q$, which corresponds to a shorter flying distance. The slopes of the curves decrease as $q$ grows, which is consistent with the theoretical results given in Theorem 4.9. The completion time for all the tasks decreases for about 8% when we change the sensing probability threshold from 0.9 to 0.8, and it further decreases about 6% if the threshold is reduced to 0.7.

Figure 4.11 shows the completion time for all the tasks $T_{max}$ v.s. the number of tasks for a UAV $N_i$. In the ITSSO and FSL schemes, we set the number of UAVs as 20, and the number of cooperative UAVs for a task as 4. In the NC scheme, the number of UAVs is also set as 20, and each task is only performed with one UAV. The completion time for all the tasks increases linearly with the number of tasks for a UAV. The slope of the ITSSO scheme is about 15% lower than that of the NC scheme due to a shorter average UAV moving distance. The completion time for all the tasks of the FSL scheme is over 50% larger than the ITSSO scheme due to the lack of sensing location optimization.

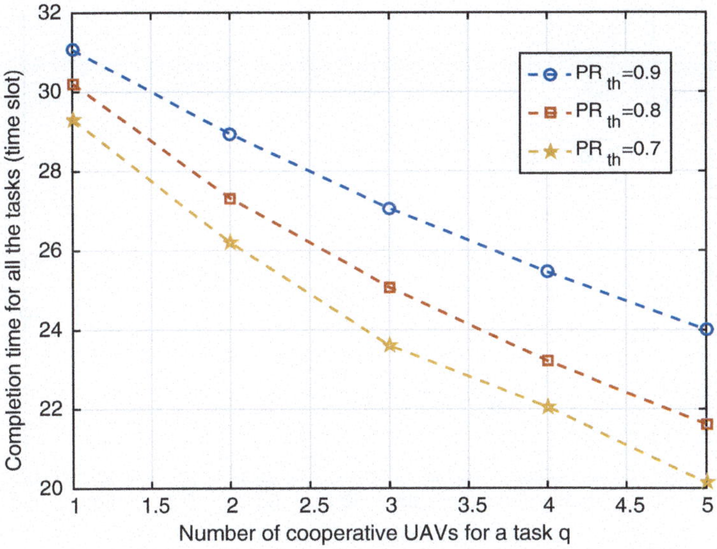


**Fig. 4.10** Number of cooperative UAVs for a task vs. completion time for all the tasks ($N = 20$, $N_i = 4$)

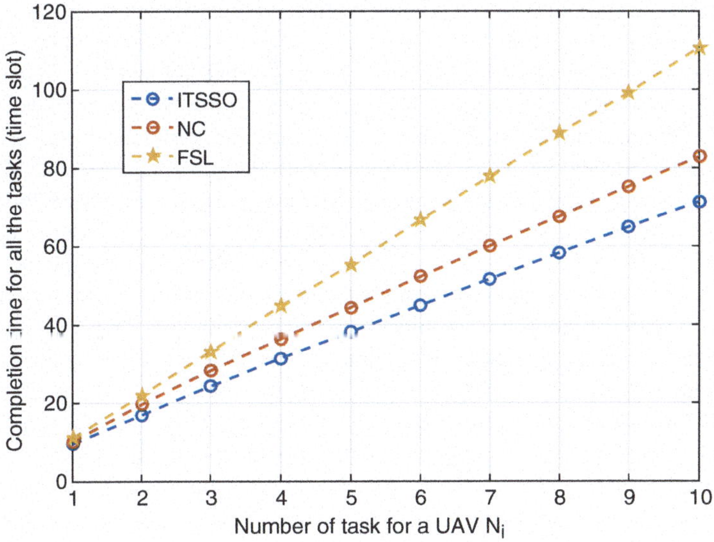


**Fig. 4.11** Number of tasks for a UAV vs. completion time for all the tasks ($M = 20$, $q = 4$)

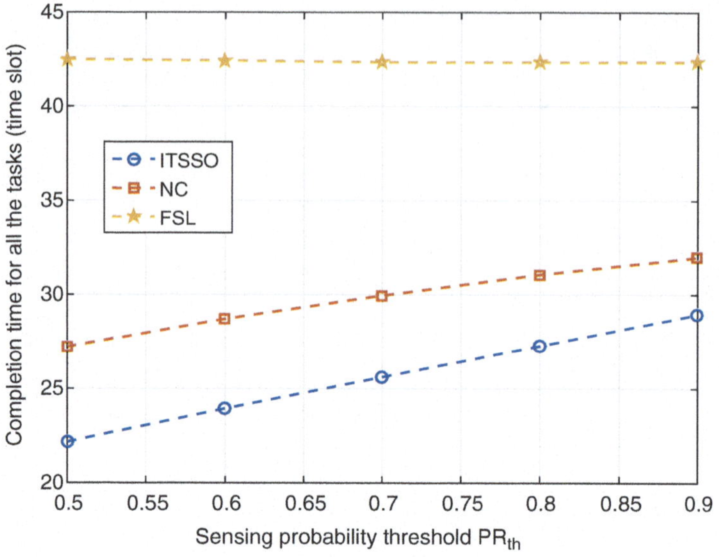


**Fig. 4.12** Sensing probability threshold vs. completion time for all the tasks ($N = 20$, $N_i = 4$, $q = 4$)

In Fig. 4.12, we plot the relation between the sensing probability threshold $PR_{th}$ and the completion time for all the tasks $T_{max}$. Here, we set the number of tasks as 20, and the number of tasks for a UAV as 4. In the ITSSO and FSL schemes, the number of cooperative UAVs for a task is set as 4. In the NC scheme, each task is performed by only one UAV. We can observe that the completion time for all the tasks of the ITSSO and NC scheme increases with the sensing probability threshold due to the change of the sensing locations. The completion time for all the tasks in the ITSSO scheme increases from 22 to 29 when the sensing probability threshold increases from 0.5 to 0.9, and the completion time for all the tasks in the NC scheme increases from 27 to 32 in the same range of sensing probability threshold. The slope is consistent with the theoretical results given in Theorem 4.10. The completion time for all the tasks in the FSL scheme is around 42, since the sensing locations are fixed in this scheme.

Figure 4.13 shows the completion time for all the tasks as a function of the sensing task size $R_s$. We set the number of tasks as 20, and the number of UAVs as 10. In the ITSSO and FSL scheme, the number of cooperative UAVs for a task is set as 4. In the NC scheme, each task is performed by only one UAV. In the ITSSO scheme, the completion time for all the tasks is around 28–30 when $R_s \le 25$ Mbps, where most of the data transmission tasks can be completed without a detour. The completion time for all the tasks starts to increase significantly when $R_s > 25$ Mbps, since the UAVs prefer to detour to the BS for data transmission. Note that the increment of the completion time for all the tasks is mainly caused by the trajectory detouring. Therefore, the completion time for all the tasks in these three schemes

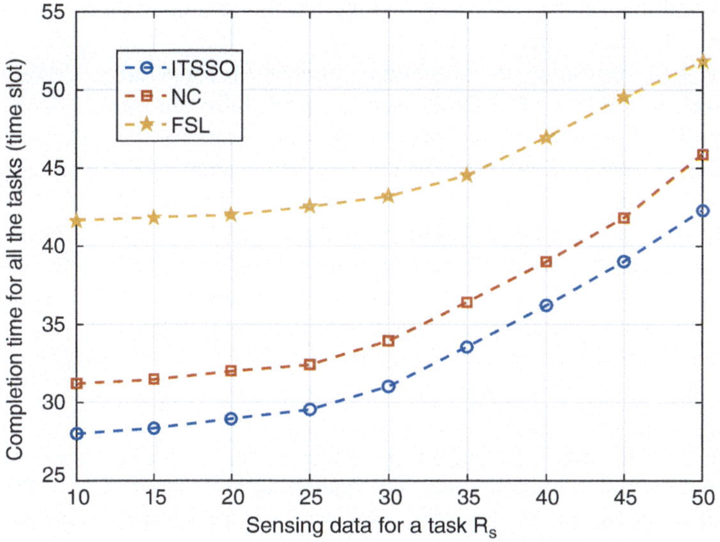


**Fig. 4.13** Sensing data size vs. completion time for all the tasks ($N = 20$, $N_i = 4$, $q = 4$)

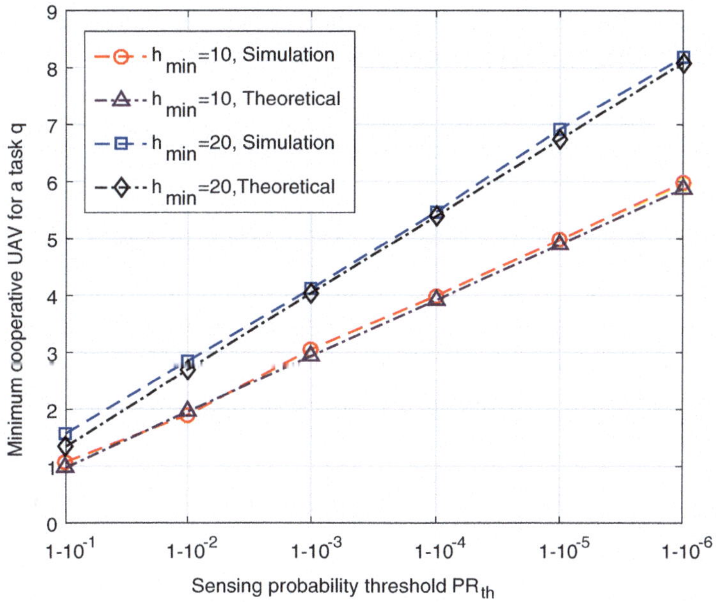


**Fig. 4.14** Sensing probability threshold vs. minimum number of cooperative UAVs for a task

increases with the sensing data size, and the difference among these three schemes decreases with a larger $R_s$.

In Fig. 4.14, we study the minimum number of cooperative UAVs required for completing a task with different sensing probability thresholds. Given a high sensing probability requirement, cooperation is necessary for the UAVs to complete a sensing task. We can observe that the minimum number of UAVs required for a task is logarithmically related to the sensing probability threshold, which is consistent with the theoretical results in (4.18) and (4.19). The difference between the simulation result and the theoretical result is less than 0.2 within the simulation range. When we set $PR_{th} = 1-10^{-1}$, the minimum number of UAVs required for a task is 1. When the sensing probability threshold is set as $PR_{th} = 1 - 10^{-6}$, at least 6 UAVs are required to complete a task. The required number of cooperative UAVs increases for about 50% when we adjust the minimum UAV altitude $h_{min}$ from 10 to 20 m.

Figure 4.15 illustrates the impact of subchannel number $K$ and task size $R_s$ on the completion time for all the tasks. The number of UAV is set as $M = 20$, and the number of cooperative UAV for a task is set as $q = 4$. It is shown that the completion time for all the tasks is strongly affected by the task size $R_s$ with $R_s \leq 30$ Mbps, and the marginal impact of $R_s$ decreases when $R_s > 30$ Mbps. Given a fixed $R_s$, the completion time for all the tasks decreases rapidly with the number of subchannels with $K \leq 4$. The ratio decreases until convergence with the increment of the number of subchannels. The simulation curves verifies the analysis of the dominated factor of the completion time for all the tasks given in Sect. 4.2.4.4.

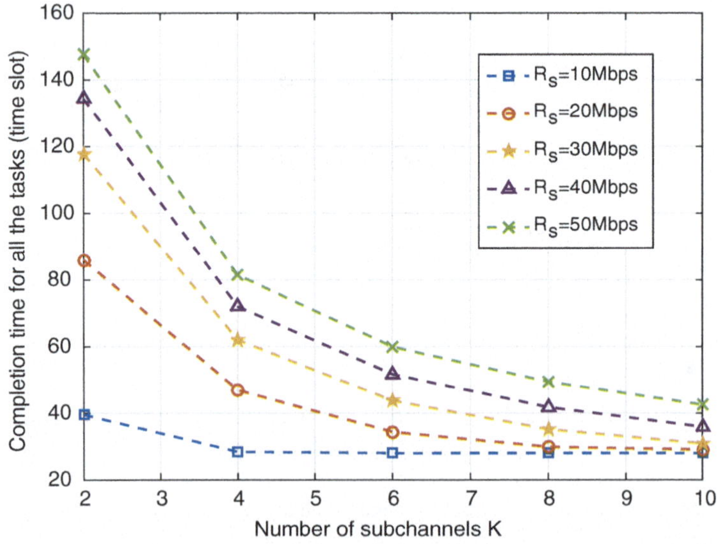


**Fig. 4.15** Number of subchannels vs. completion time for all the tasks ($q = 4$, $M = 20$)

**Table 4.3** Impact of UAV number $M$ on computational time ($N = 20$, $N_i = 4$)

| Number of UAVs $M$ | 20 | 40 | 60 | 80 | 100 |
|---|---|---|---|---|---|
| Computation time (s) | 0.0311 | 0.0318 | 0.0347 | 0.0466 | 0.0911 |

**Table 4.4** Impact of task number $N$ on computational time ($M = 20$, $q = 4$)

| Number of tasks $N$ | 20 | 40 | 60 | 80 | 100 |
|---|---|---|---|---|---|
| Computation time (s) | 0.0311 | 0.0320 | 0.0391 | 0.1023 | 0.6024 |

Table 4.3 shows the impact of the number of UAVs $M$ on the computational time with the proposed algorithm. The number of tasks $N$ is given as 20, and the number of tasks for each UAV $N_i$ is given as 4. The number of cooperative UAVs for each task $q$ is linearly related with the number of UAVs $M$. Therefore, the computational time should increase quadratically with the number of UAVs $N$ according to the theoretical algorithm complexity derived in Theorem 4.8, i.e., $O(N^2 \times q \times N_i \times M)$. Simulation results suit the trend of quadratical increment well, which is in accordance with the analysis.

Table 4.4 shows the impact of the number of tasks $N$ on the computational time with the proposed algorithm. Given number of UAVs $M$ as 20, and the number of the UAVs for each task $q$ as 4, the number of tasks for each UAV $N_i$ increases linearly with the number of tasks $N$. Therefore, the computational time should increase cubically with the number of tasks $N$ according to the theoretical algorithm complexity derived in Theorem 4.8, i.e., $O(N^2 \times q \times N_i \times M)$. The result of the simulation is consistent with the theoretical results.

### 4.2.6 Summary

In this section, we have studied a single cell UAV sensing network where multiple UAVs perform cooperative sensing and transmission. We first have proposed a sense-and-send protocol to facilitate the cooperation, and have formulated a joint trajectory, sensing location, and UAV scheduling optimization problem to minimize the completion time for all the tasks. To solve the NP-hard problem, we have decoupled it into three subproblems: trajectory optimization, sensing location optimization, and UAV scheduling, and have proposed an iterative algorithm to solve it. We then have analyzed the system performance, from which we can infer that the UAV cooperation reduces the completion time for all the tasks, and the marginal gain becomes smaller as the number of cooperative UAVs grows. Simulation results have shown that the completion time for all the tasks in the proposed ITSSO scheme is 15% less than the NC scheme, and over 50% less than the FSL scheme. The transmission resource is a dominated factor on the completion time for all the tasks when it is at a low level. The sensing data size is a dominated factor of the completion time for all the tasks when the size of the sensing task and the transmission resource are both at a high level, or both at a low level.

## 4.3 UAV-to-X Communications

In the cellular Internet of UAVs, the sensory data needs to be transmitted to the server for further processing, thereby posing high uplink rate requirement on the UAV communication network. In this section, we study a single cell cellular Internet of UAVs with a number of cellular users (CUs) and UAVs, where each UAV moves along a pre-determined trajectory to collect data, and then uploads these data to the BS. However, some UAVs may locate at the cell edge, and the SNR of their communication links to the BS is low. To provide a satisfactory data rate, we enable these UAVs to transmit the sensory data to the UAVs with high SNR for the communication link to the BS as relay. The relaying UAVs save the received data in their caches and upload the data to the BS in the following time slots as described in [28]. Specifically, the UAV transmissions can be supported by two basic modes, namely UAV-to-network (U2N) and UAV-to-UAV (U2U) transmissions. The overlay U2N transmission offers direct link from UAVs with high SNR to the BS, and thus provides a high data rate [29, 30]. In U2U transmission, a UAV with low SNR for the U2N link can set up direct communication links to the high U2N SNR UAVs bypassing the network infrastructure and share the spectrum with the U2N and CU transmissions, which provides a spectrum-efficient method to support the data relaying process [31].

Due to the high mobility and long transmission distance of the sensing UAVs, it is not trivial to address the following issues. Firstly, since the U2U transmissions underlay the spectrum resources of the U2N and CU transmissions, the U2N and CU transmissions may be interfered by the U2U transmissions when sharing the same subchannel. Correspondingly, the U2U transmissions are also interfered by the U2N, CU, and U2U links on the same subchannel. Moreover, different channel models are utilized for the U2N, U2U, and CU transmissions due to the different characteristics of air-to-ground, air-to-air, and ground-to-ground communications. Therefore, an *efficient spectrum allocation algorithm* is required to manage the mutual interference. Secondly, to complete the data collection of the sensing tasks given time requirements, the *UAV speed optimization* is necessary. Thirdly, to avoid the data loss and provide a relatively high data rate for the UAVs with low SNR for the link to the BS, an *efficient communication method* is essential. In summary, the resource allocation schemes, UAV speed, and UAV transmission protocol should be properly designed to support the UAV-to-X communications.

The rest of this section is organized as follows. In Sect. 4.3.1, we present the system model of the UAV sensing network. A cooperative UAV sense-and-send protocol is proposed in Sect. 4.3.2 for the data collection and UAV-to-X communications. In Sect. 4.3.3, we formulate the uplink sum-rate maximization problem by optimizing the subchannel allocation and UAV speed jointly. The ISASOA is proposed in Sect. 4.3.4, followed by the corresponding analysis. Simulation results are presented in Sect. 4.3.5, and finally we summarize the section in Sect. 4.3.6.

## *4.3.1 System Model*

In this subsection, we first describe the working scenario, and then introduce the data transmission of this network. Finally, we present the channel models for U2N, U2U, and CU transmissions, respectively.

### 4.3.1.1 Scenario Description

We consider a single cell cellular Internet of UAVs as shown in Fig. 4.16 [16], which consists of one BS, $M$ CUs, denoted by $\mathscr{M} = \{1, 2, \cdots, M\}$, and $N$ UAVs, denoted by $\mathscr{N} = \{1, 2, \cdots, N\}$. The UAVs collect various required data with their sensors in each time slot, and the data will be transmitted to the BS for further processing. In each time slot, the UAVs first perform UAV sensing, and then perform data transmission. The length of time for UAV sensing and data transmission is given in each time slot. We assume that each UAV flies along a pre-determined trajectory during the sensing and transmission process. The speeds of the UAVs in each time slot are not given, but all the UAVs are required to arrive at the endpoints of their trajectories within a number of time slots for timely sensing and transmission. To provide a high data transmission rate for all the UAVs, we

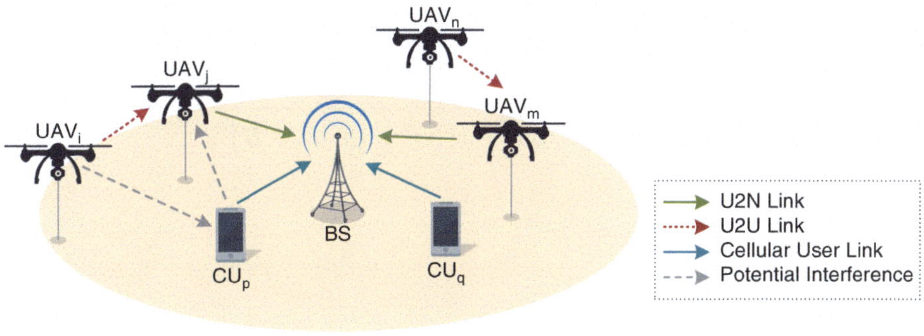


**Fig. 4.16** System model of U2X communications

distinguish the UAVs with different QoS requirements for the U2N links into two transmission modes, namely U2N transmission and U2U transmission. The UAVs in the U2N mode transmit overlaying the cellular ones, while the UAVs in the U2N mode can transmit underlaying the U2N and CU transmissions, i.e., they can reuse the spectrum resources occupied by the U2N and CU transmissions. The criterion of adopting U2N or U2U modes will be elaborated on in Sect. 4.3.1.2.

We denote the location of UAV $i$ in time slot $t$ by $\mathbf{l}_i(t) = (x_i(t), y_i(t), h_i(t))$, and the location of the BS by $(0, 0, H)$. Each UAV moves along a pre-determined trajectory. Let $v_i(t)$ be the speed of UAV $i$ in time slot $t$. The location of UAV $i$ in time slot $t+1$ is given as $\mathbf{l}_i(t+1) = \mathbf{l}_i(t) + v_i(t) \cdot \boldsymbol{\omega}_i(t)$, where $\boldsymbol{\omega}_i(t)$ is the trajectory direction of UAV $i$ in time slot $t$. Due to the mechanical limitation, the speed of a UAV is no more than $v_{max}$.[4] Let $L_i$ be the length of UAV $i$'s trajectory. With proper transmission rate requirements, the UAVs are capable to upload the sensory data to the BS with low latency. Therefore, the task completion time of a UAV can be defined as the time that it costs to complete its moving along the trajectory, which is determined by its speed in each time slot. For timely data collection, the task completion time of each UAV is required to be no more than $T$ time slots, i.e., $\sum_{t=1}^{T} v_i(t) \geq L_i, \forall i \in \mathcal{N}$. In time slot $t$, the distance between UAV $i$ and UAV $j$ is shown as

$$d_{i,j}(t) = \sqrt{\left(x_i(t) - x_j(t)\right)^2 + \left(y_i(t) - y_j(t)\right)^2 + \left(h_i(t) - h_j(t)\right)^2}, \tag{4.41}$$

and the distance between UAV $i$ and the BS is expressed as

$$d_{i,BS}(t) = \sqrt{\left(x_i(t)\right)^2 + \left(y_i(t)\right)^2 + \left(h_i(t) - H\right)^2}. \tag{4.42}$$

The location of CU $i$ is given as $\left(x_i^c, y_i^c, h_i^c\right)$. We assume that the locations of the CUs are fixed in different time slots, as the mobility of the CUs is much lower than that of the UAVs. Therefore, the distance between CU $i$ and UAV $j$ can be denoted by

$$d_{i,j}^c(t) = \sqrt{\left(x_i^c(t) - x_j(t)\right)^2 + \left(y_i^c(t) - y_j(t)\right)^2 + \left(h_i^c(t) - h_j(t)\right)^2}, \tag{4.43}$$

and the distance between UAV $i$ and BS can be shown as

$$d_{i,BS}^c(t) = \sqrt{x_i^c(t)^2 + y_i^c(t)^2 + \left(h_i^c(t) - H\right)^2}. \tag{4.44}$$

[4]We consider the UAVs as rotary wing UAVs which can hover in the air for some time slots. The rotary wing UAVs can move with the speed of $[0, v_{max}]$ in any time slot.

#### 4.3.1.2 Data Transmission

In this part, we give a brief introduction to the data transmission of this network. There are two UAV transmission modes in this network: U2N and U2U modes. A UAV may transmit in either the U2N or U2U mode in one time slot. The criterion of adopting U2N or U2U mode is given below.

1. **U2N Mode**: A UAV with the high SNR for the link to the BS performs U2N transmission in the network. It uploads its collected data to the BS directly over the assigned subchannel.
2. **U2U Mode**: A UAV with the low SNR for the link to the BS performs U2U communication to transmit the collected data to a UAV in the U2N transmission mode.

The detailed method for the U2N or U2U mode selection will be described in Sect. 4.3.2. Let $\mathscr{N}_h(t) = \{1, 2, \cdots, N_h(t)\}$ and $\mathscr{N}_l(t) = \{1, 2, \cdots, N_l(t)\}$ be the set of UAVs in U2N and U2U modes in time slot $t$, respectively, with $\mathscr{N} = \mathscr{N}_h(t) \cup \mathscr{N}_l(t)$. For the UAVs in $\mathscr{N}_h(t)$, they send the data to the BS by U2N transmissions. For the UAVs in $\mathscr{N}_l(t)$, the SNR of the direct communication links is low, and thus these UAVs are difficult to provide high data rates to support timely data upload via U2N modes. Therefore, the UAVs send the collected data to the neighboring UAVs with the high SNR for the U2N link via U2U transmissions, and the data will be sent to the BS later by the relaying UAVs.

The transmission bandwidth of this network is divided into $K$ orthogonal subchannels, denoted by $\mathscr{K} = \{1, 2, \cdots, K\}$. It is worthwhile to mention that a single UAV can perform U2N transmission and U2U reception over different subchannels simultaneously. For the sake of transmission quality, we assume that a subchannel can serve at most one U2N or CU link, but multiple U2U links in one time slot. In addition, to guarantee fairness among the users, we also assume that each transmission link can be allocated with no more than $\chi_{max}$ subchannels. In time slot $t$, we define a binary U2N and CU subchannel pairing matrix $\boldsymbol{\Phi}(t) = [\phi_{i,k}(t)]_{(N_h+M)\times K}$, and a binary U2U subchannel pairing matrix $\boldsymbol{\Psi}(t) = [\psi_{i,k}(t)]_{N_l\times K}$, to describe the resource allocation for CU, U2N, and U2U transmissions, respectively. For $i \leq N_h$, $\phi_{i,k}(t) = 1$ when subchannel $k$ is assigned to UAV $i$ for U2N transmission, otherwise $\phi_{i,k}(t) = 0$. For $i > N_h$, $\phi_{i,k}(t) = 1$ when subchannel $k$ is assigned to CU $i - N$ for CU transmission, otherwise $\phi_{i,k}(t) = 0$. Likewise, the value of $\psi_{i,k}(t) = 1$ when subchannel $k$ is assigned to UAV $i$ in the U2U mode, otherwise $\psi_{i,k}(t) = 0$.

We denote $\xi_{i,j}(t) = 1$ when UAV $i$ performs U2U transmission with UAV $j$ in time slot $t$, and $\xi_{i,j}(t) = 0$ otherwise. In order to avoid the high communication latency for the UAVs, the data rate of each U2U communication link should be no less $R_0$, i.e., $\sum_{k=1}^{K} \psi_{i,k}(t) R_{i,j}^{k}(t) \geq R_0, \forall i, j \in \mathscr{N}, \xi_{i,j} = 1$.

#### 4.3.1.3 Channel Model

In this subsection, we introduce the channel model in this network. The channel models of the U2N, CU, and U2U transmissions are different, due to the different characteristics in the LoS probability and the elevation angel, which will be introduced as follows, respectively.

1. *U2N Channel Model*: We use the air-to-ground propagation model which is proposed in [11, 32, 33] for the U2N transmission. In time slot $t$, the LoS and NLoS pathloss from UAV $i$ to the BS are given by

$$PL_{LoS,i}(t) = L_{FS,i}(t) + 20\log(d_{i,BS}(t)) + \eta_{LoS}, \tag{4.45}$$

$$PL_{NLoS,i}(t) = L_{FS,i}(t) + 20\log(d_{i,BS}(t)) + \eta_{NLoS}, \tag{4.46}$$

where $L_{FS,i}(t)$ is the free space pathloss given by $L_{FS,i}(t) = 20\log(f) + 20\log(\frac{4\pi}{c})$, and $f$ is the system carrier frequency. $\eta_{LoS}$ and $\eta_{NLoS}$ are additional attenuation factors due to the LoS and NLoS connections. Considering the antennas on UAVs and the BS placed vertically, the probability of LoS connection is given by

$$P_{LoS,i}(t) = \frac{1}{1 + a\exp(-b(\theta_i(t) - a))}, \tag{4.47}$$

where $a$ and $b$ are constants which depend on the environment, and the elevation angle $\theta_i(t) = \sin^{-1}((h_i(t) - H)/d_{i,BS}(t))$. The average pathloss in dB can then be expressed as

$$PL_{avg,i}(t) = P_{LoS,i}(t) \times PL_{Los,i}(t) + P_{NLoS,i}(t) \times PL_{NLoS,i}(t), \tag{4.48}$$

where $P_{NLoS}(t) = 1 - P_{LoS}(t)$. The average received power of BS from UAV $i$ over its paired subchannel $k$ is given by

$$P^k_{i,BS}(t) = \frac{P_U}{10^{PL_{avg,i}(t)/10}}, \tag{4.49}$$

where $P_U$ is the transmit power of a UAV or CU over one subchannel. Since each subchannel can be assigned to at most one U2N or CU link, the interference to the U2N links only comes from the U2U links due to spectrum sharing. When UAV $i$ in the U2N mode over subchannel $k$, the U2U interference is expressed as

$$I_{k,U2U}(t) = \sum_{j=1}^{N_I} \psi_{j,k}(t) P^k_{j,BS}(t). \tag{4.50}$$

Therefore, the signal to interference plus noise ratio (SINR) of the BS over subchannel $k$ is given by

$$\gamma_{i,BS}^{k}(t) = \frac{P_{i,BS}^{k}(t)}{\sigma^2 + I_{k,U2U}(t)}, \tag{4.51}$$

where $\sigma^2$ is the variance of AWGN with zero mean. The data rate that BS receives from UAV $i$ over subchannel $k$ is shown as

$$R_{i,BS}^{k}(t) = \log_2(1 + \gamma_{i,BS}^{k}(t)). \tag{4.52}$$

2. *CU Channel Model*: We utilize the macrocell pathloss model as proposed in [34]. For CU $i$, the pathloss in dB can be expressed by

$$PL_{i,C}^{k}(t) = -55.9 + 38\log(d_{i,BS}^{c}(t)) + (24.5 + 1.5f/925)\log(f). \tag{4.53}$$

When CU $i$ transmits signals to BS, the received power is expressed as

$$P_{i,C}^{k}(t) = \frac{P_U}{10^{PL_{i,C}^{k}(t)/10}}. \tag{4.54}$$

We denote the set of UAVs that share subchannel $k$ with CU $i$ by $U_i = \{m|\psi_{m,k}(t) = 1, \forall m \in \mathcal{N}_e\}$, and the received power at the BS over subchannel $k$ is shown as

$$y_{i,j}^{k}(t) = \sqrt{P_{i,C}^{k}(t)} + \sum_{m \in U_i} \sqrt{P_{m,BS}^{k}(t)} + n_j^k(t), \tag{4.55}$$

where $n_j^k(t)$ is the AWGN with zero mean and $\sigma^2$ variance. Therefore, the received signal at the BS over subchannel $k$ can be given by

$$\gamma_{i,BS}^{k}(t) = \frac{P_{i,C}^{k}(t)}{\sigma^2 + I_{k,U2U}(t)}, \tag{4.56}$$

where $I_{k,U2U}(t) = \sum_{j=1}^{N} \psi_{j,k}(t) P_{j,BS}^{k}(t)$ is the U2U interference. The data rate for CU $i$ over subchannel $k$ is expressed as

$$R_{i,BS}^{k}(t) = \log_2(1 + \gamma_{i,BS}^{k}(t)). \tag{4.57}$$

3. *U2U Channel Model*: For U2U communication, free space channel model is utilized. When UAV $i$ transmits signals to UAV $j$ over subchannel $k$, the received power at UAV $j$ from UAV $i$ is expressed as

$$P_{i,j}^{k}(t) = P_U G(d_{i,j}(t))^{-\alpha}, \tag{4.58}$$

where $G$ is the constant power gains factor introduced by amplifier and antenna, and $(d_{i,j}(t))^{-\alpha}$ is the pathloss. Define the set of UAVs and CUs that share subchannel $k$ with UAV $i$ as $W_i = \{m|\psi_{m,k}(t) = 1, \forall m \in \mathcal{N}_e \backslash i\} \cup \{m|\phi_{m,k}(t) = 1\}$. The received signal at UAV $j$ over subchannel $k$ is then given by

$$y_{i,j}^k(t) = \sqrt{P_{i,j}^k(t)} + \sum_{m \in W_i} \sqrt{P_{m,j}^k(t)} + n_j^k(t), \tag{4.59}$$

where $P_m^k(t)$ is the received power at UAV $j$ from the UAVs and CUs in $W_i$, and $n_j^k(t)$ is the AWGN with zero mean and $\sigma^2$ variance. The interference from UAV $m$ to UAV $j$ over subchannel $k$ is shown as

$$I_{m,UAV}^k(t) = (\phi_{m,k}(t) + \psi_{m,k}(t)) P_U G (d_{m,j}(t))^{-\alpha}. \tag{4.60}$$

According to the channel reciprocity, the interference from CU $m$ to UAV $j$ over subchannel $k$ can be expressed as

$$I_{m,C}^k(t) = \phi_{m,k}(t) \frac{P_U}{10^{PL_{avg,j}^m(t)/10}}, \tag{4.61}$$

where $PL_{avg,j}^m(t)$ is the average pathloss from UAV $j$ to CU $m$, which can be derived from Eqs. (4.45)–(4.48). The SINR at UAV $j$ over subchannel $k$ is shown as

$$\gamma_{i,j}^k(t) = \frac{P_U G (d_{i,j}(t))^{-\alpha}}{\sigma^2 + \sum\limits_{m=1, m \neq i}^{N_l + N_h} I_{m,UAV}^k(t) + \sum\limits_{m=1}^{M} I_{m,C}^k(t)}. \tag{4.62}$$

When UAV $i$ transmits its data to UAV $j$ over subchannel $k$ via U2U transmission, the data rate is given by

$$R_{i,j}^k(t) = \log_2(1 + \gamma_{i,j}^k(t)). \tag{4.63}$$

### 4.3.2 *Cooperative UAV Sense-and-Send Protocol*

In this subsection, we propose a cooperative UAV sense-and-send protocol that supports the UAV data collections and U2X transmissions in this network. As illustrated in Fig. 4.17, in each time slot, the UAVs first collect the sensory data of their tasks. They then send beacons to the BS over the control channel, and the BS selects the transmission modes for the UAVs according to the received SNR. Afterwards, the BS performs U2U pairing, subchannel allocation, and UAV speed optimization for the UAVs in the network, and sends the results to the UAVs. After receiving the results, the UAVs establish the transmission links, and perform U2N

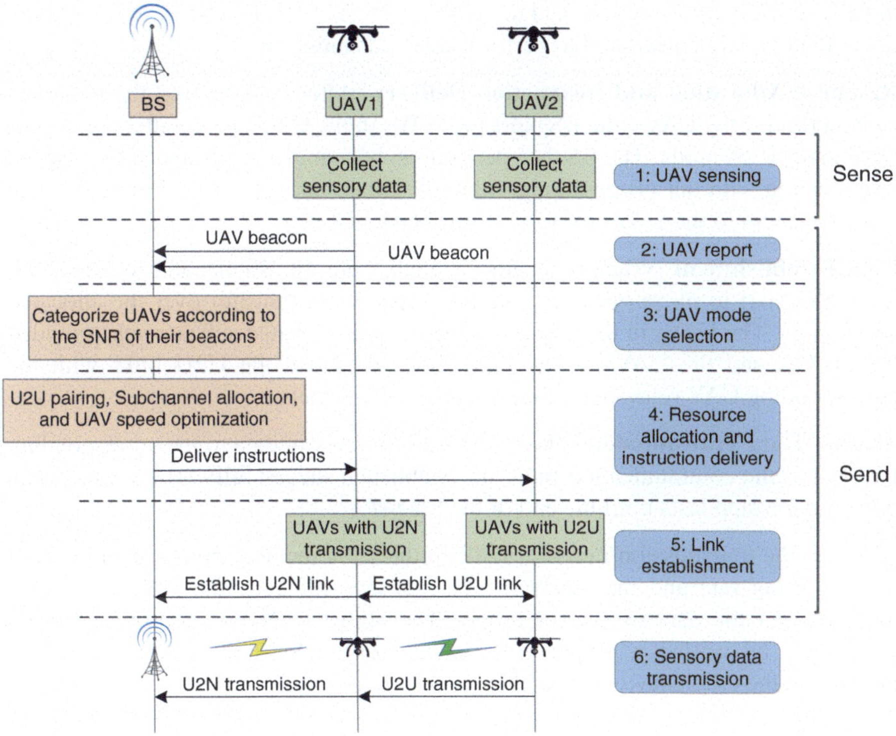


**Fig. 4.17** Cooperative UAV sense-and-send protocol

and U2U transmissions according to the arrangement of the BS. To better describe the protocol, we divide each time slot into six steps: UAV sensing, UAV report, UAV mode selection, resource allocation and instruction delivery, link establishment, and sensory data transmission, and introduce them in details in the following.

**UAV Sensing** In the UAV sensing step, the UAVs perform sensing and save the collected data in their caches. The communication module is turned off in the UAV sensing step.

**UAV Report** After the UAV sensing step, the UAVs stop data collection and send beacons to the BS. The beacon of each UAV contains its ID and current location, and is sent to the BS over the control channel in a time-division manner.

**UAV Mode Selection** When receiving the beacons of the UAVs, the BS categorizes the UAVs into U2N and U2U modes according to the received SNR. A SNR threshold $\gamma_{th}$ is given to distinguish the UAVs that transmit in the U2N and U2U modes.[5] The UAVs with the SNR for U2N links being larger than $\gamma_{th}$ are considered

[5]The value of $\gamma_{th}$ is set according to the QoS in the specific network.

to perform U2N transmission, and the UAVs with the SNR for the U2N links being lower than $\gamma_{th}$ are considered to perform U2U transmission.

**Resource Allocation and Instruction Delivery** After categorizing the transmission modes for the UAVs, the BS pairs the UAVs in the U2U mode with their closest UAV in the U2N mode. The BS then performs subchannel allocation and UAV speed optimization with our proposed algorithm described in Sect. 4.3.4. Afterwards, the results are sent to the UAVs over the control channel.

**Link Establishment** When the control signals from the BS are sent to the UAVs, the UAVs start to move with the optimized speed and transmit over the allocated subchannel. The UAVs in the U2N mode access the allocated subchannels provided by the BS, and the UAVs in the U2U mode establish the U2U links with the corresponding UAV relays over the allocated subchannels.

**Sensory Data Transmission** The UAVs start to transmit data to the corresponding target after the communication links are established successfully. The sensory data transmission step lasts until the end of the time slot.

When the transmission rate of a UAV in the U2N mode is higher than the sum of its sensing rate and the received U2U transmission rate, the UAV is capable to upload all the data to the BS timely. The above U2N rate constraint can be guaranteed by setting a proper UAV categorization SNR threshold $\gamma_{th}$, which guarantees the efficiency of this network.

## *4.3.3 Problem Formulation*

In this subsection, we first formulate the joint subchannel allocation and UAV speed optimization problem, and prove that the optimization problem is NP-hard, which cannot be solved directly within polynomial time. Therefore, in the next part, we decouple it into three subproblems, and elaborate them separately.

### 4.3.3.1 Joint Subchannel Allocation and UAV Speed Optimization Problem Formulation

Since all the data collected by the UAVs needs to be sent to the BS, the uplink sum-rate of this network is one key metric to evaluate the performance of this network.[6] In time slot $t$, we denote the set of UAVs that have not completed the task along their trajectories by $\lambda(t)$. We aim to maximize the uplink sum-rate of the UAVs in

[6]The uplink sum-rate is the sum of U2N and CU transmission rate, and U2U transmission is not included in the uplink sum-rate. Therefore, the objective function does not contain the U2U rate. Instead, we set a minimum threshold for each U2U link to guarantee the success of the U2U transmissions.

$\lambda(t)$ and the CUs by optimizing the subchannel allocation and UAV speed variables $\boldsymbol{\Phi}(t)$, $\boldsymbol{\Psi}(t)$, and $v_i(t)$. The joint subchannel allocation and UAV speed optimization problem can be formulated as follows:

$$\max_{\substack{\{v_i(t)\},\{\boldsymbol{\Phi}(t)\},\\ \{\boldsymbol{\Psi}(t)\}}} \sum_{k=1}^{K} \sum_{\substack{i=1\\ i\in\lambda(t)}}^{N_h+M} \phi_{i,k}(t) R_{i,BS}^{k}(t), \tag{4.64a}$$

$$\textbf{\textit{s.t.}} \sum_{k=1}^{K} \psi_{i,k}(t) R_{i,j}^{k}(t) \geq R_0, \forall i, j \in \mathscr{N}, \xi_{i,j} = 1, \tag{4.64b}$$

$$v_i(t) \leq v_{max}, \forall i \in \mathscr{N}, \tag{4.64c}$$

$$\sum_{t=1}^{T} v_i(t) \geq L_i, \forall i \in \mathscr{N}, \tag{4.64d}$$

$$\sum_{i=1}^{N_h+M} \phi_{i,k}(t) \leq 1, \forall k \in \mathscr{K}, \tag{4.64e}$$

$$\sum_{k=1}^{K} \phi_{i,k}(t) \leq \chi_{max}, \forall i \in \mathscr{N}_h(t) \cup \mathscr{M}, \tag{4.64f}$$

$$\sum_{k=1}^{K} \psi_{i,k}(t) \leq \chi_{max}, \forall i \in \mathscr{N}_l(t), \tag{4.64g}$$

$$\phi_{i,k}(t), \psi_{i,k}(t) \in \{0, 1\}, \forall i \in \mathscr{N} \cup \mathscr{M}, k \in \mathscr{K}. \tag{4.64h}$$

The minimum U2U transmission rate satisfies constraint (4.64b). Equation (4.64c) is the maximum speed constraint for the UAVs, and (4.64d) shows that the task completion time of each UAV is no more than $T$ time slots. Constraint (4.64e) implies that each subchannel can be allocated to at most one UAV in the U2N mode or CU. Each UAV and CU can be paired with at most $\chi_{max}$ subchannels, which is given in constraints (4.64f) and (4.64g). In the following theorem, we will prove that optimization problem (4.64) is NP-hard.

**Theorem 4.12** *Problem (4.64) is NP-hard.*

***Proof*** We prove that problem (4.64) is NP-hard even when we do not perform UAV speed optimization. We construct an instance of problem (4.64) where each subchannel can only serve no more than one U2U link and one U2N or CU link simultaneously. Let $\mathscr{N}_c$, $\mathscr{N}_e$, and $\mathscr{K}$ be three disjoint sets of UAVs in the U2N mode and CUs, UAVs in the U2U mode, and subchannels, respectively, with $|\mathscr{N}_c| = N_h$, $|\mathscr{N}_e| = N_l$, and $|\mathscr{K}| =$ K. Set $\mathscr{N}_c$, $\mathscr{N}_e$, and $\mathscr{K}$ satisfy $\mathscr{N}_c \cap \mathscr{N}_e = \varnothing$, $\mathscr{N}_c \cap \mathscr{K} = \varnothing$, and $\mathscr{N}_e \cap \mathscr{K} = \varnothing$. Let $\mathscr{P}$ be a collection of ordered triples $\mathscr{P} \subseteq \mathscr{N}_c \times \mathscr{N}_e \times \mathscr{K}$, where each element in $\mathscr{P}$ consists a CU/UAV that perform U2N transmission, a

UAV that perform U2U transmission, and a subchannel, i.e., $P_i = \left(N_{c,i}, N_{e,i}, K_i\right) \in \mathscr{P}$. To be convenient, we set $L = \min\{N_h, N_l, K\}$. There exists $\mathscr{P}' \subseteq \mathscr{P}$ that holds: (1) $\left|\mathscr{P}'\right| = L$; (2) for any two distinct triples $\left(N_{c,i}, N_{e,i}, K_i\right) \in \mathscr{P}'$ and $\left(N_{c,j}, N_{e,j}, K_j\right) \in \mathscr{P}'$, we have $i \neq j$. Therefore, $\mathscr{P}'$ is a three dimension matching (3-DM). Since 3-DM problem has been proved to be NP-complete in [35], the constructed instance of problem is also NP-complete. Thus, the problem in (4.64) is NP-hard [36]. □

#### 4.3.3.2 Problem Decomposition

Since problem (4.64) is NP-hard, to tackle this problem efficiently, we decouple problem (4.64) into three subproblems, i.e., U2N and CU subchannel allocation, U2U subchannel allocation, and UAV speed optimization subproblems. In the U2N and CU subchannel allocation subproblem, the U2U subchannel matching matrix $\boldsymbol{\Psi}(t)$ and the UAV speed $\{v_i(t)\}$ are considered to be fixed. Therefore, the U2N and CU subchannel allocation subproblem is written as

$$\max_{\boldsymbol{\Phi}(t)} \sum_{k=1}^{K} \sum_{\substack{i=1 \\ i \in \lambda(t)}}^{N_h+M} \phi_{i,k}(t) R_{i,BS}^{k}(t), \tag{4.65a}$$

$$s.t. \sum_{i=1}^{N_h+M} \phi_{i,k}(t) \leq 1, \forall k \in \mathscr{K}, \tag{4.65b}$$

$$\sum_{k=1}^{K} \phi_{i,k}(t) \leq \chi_{max}, \forall i \in \mathscr{N}_h(t) \cup \mathscr{M}, \tag{4.65c}$$

$$\phi_{i,k}(t) \in \{0, 1\}, \forall i \in \mathscr{N}_h(t) \cup \mathscr{M}, k \in \mathscr{K}. \tag{4.65d}$$

Given the U2N and CU subchannel pairing matrix $\boldsymbol{\Phi}(t)$ and the UAV speed $\{v_i(t)\}$, the U2U subchannel allocation subproblem can be written as

$$\max_{\boldsymbol{\Psi}(t)} \sum_{k=1}^{K} \sum_{\substack{i=1 \\ i \in \lambda(t)}}^{N_h+M} \phi_{i,k}(t) R_{i,BS}^{k}(t), \tag{4.66a}$$

$$s.t. \sum_{k=1}^{K} \psi_{i,k}(t) R_{i,j}^{k}(t) \geq R_0, \forall i, j \in \mathscr{N}_l(t), \xi_{i,j} = 1, \tag{4.66b}$$

$$\sum_{k=1}^{K} \psi_{i,k}(t) \leq \chi_{max}, \forall i \in \mathscr{N}_l(t), \tag{4.66c}$$

$$\psi_{i,k}(t) \in \{0, 1\}, \forall i \in \mathscr{N}_l(t), k \in \mathscr{K}. \tag{4.66d}$$

Similarly, when the subchannel pairing matrices $\boldsymbol{\Phi}(t)$ and $\boldsymbol{\Psi}(t)$ are given, the UAV speed optimization subproblem can be expressed by

$$\max_{\{v_i(t)\}} \sum_{k=1}^{K} \sum_{\substack{i=1 \\ i \in \lambda(t)}}^{N_h+M} \phi_{i,k}(t) R_{i,BS}^k(t), \tag{4.67a}$$

$$\sum_{k=1}^{K} \psi_{i,k}(t) R_{i,j}^k(t) \geq R_0, \forall i, j \in \mathcal{N}, \xi_{i,j} = 1, \tag{4.67b}$$

$$v_i(t) \leq v_{max}, \forall i \in \mathcal{N}, \tag{4.67c}$$

$$\sum_{t=1}^{T} v_i(t) \geq L_i, \forall i \in \mathcal{N}. \tag{4.67d}$$

## 4.3.4 Joint Subchannel Allocation and UAV Speed Optimization

In this subsection, we propose an effective method, i.e., ISASOA to obtain a sub-optimal solution of problem (4.64) by solving its three subproblems (4.65), (4.66), and (4.67) iteratively. The U2N and CU subchannel allocation subproblem (4.65) can be relaxed to a standard linear programming problem, which can be solved by existing convex techniques, for example, CVX. We then utilize the branch-and-bound method to solve the non-convex U2U subchannel allocation subproblem (4.66). For the UAV speed optimization subproblem (4.67), we discuss the feasible region and convert it into a convex problem, which can be solved by existing convex techniques. Iterations of solving the three subproblems are performed until the objective function converges to a constant. In the following, we first elaborate on the algorithms of solving the three subproblems, respectively. Afterwards, we will provide the ISASOA, and discuss its convergence and complexity.

### 4.3.4.1 U2N and CU Subchannel Allocation Algorithm

In this subsection, we give a detailed description of the U2N and CU subchannel allocation algorithm. As shown in Sect. 4.3.3.2, the decoupled subproblem (4.65) is an integer programming problem. To make the problem more tractable, we relax the variables $\boldsymbol{\Phi}(t)$ into continuous values, and the relaxed problem is expressed as

$$\max_{\boldsymbol{\Phi}(t)} \sum_{k=1}^{K} \sum_{\substack{i=1 \\ i\in\lambda(t)}}^{N_h+M} \phi_{i,k}(t) R^k_{i,BS}(t), \tag{4.68a}$$

$$s.t. \sum_{i=1}^{N_h+M} \phi_{i,k}(t) \leq 1, \forall k \in \mathcal{K}, \tag{4.68b}$$

$$\sum_{k=1}^{K} \phi_{i,k}(t) \leq \chi_{max}, \forall i \in \mathcal{N}_h(t) \cup \mathcal{M}, \tag{4.68c}$$

$$0 \leq \phi_{i,k}(t) \leq 1, \forall i \in \mathcal{N}_h(t) \cup \mathcal{M}, k \in \mathcal{K}. \tag{4.68d}$$

When we substitute (4.51), (4.52), (4.56), and (4.57) into (4.68a), it can be observed that the pairing matrix $\boldsymbol{\Phi}(t)$ is not relevant with $R^k_{i,BS}(t)$. Therefore, $R^k_{i,BS}(t)$ is fixed in this subproblem. Note that function (4.68a) is linear with respect to the optimization variables $\boldsymbol{\Phi}(t)$, and Eqs. (4.68b), (4.68c), and (4.68d) are all linear. Thus, problem (4.68) is a standard linear programming problem, which can be solved efficiently by utilizing the existing optimization techniques such as CVX [37]. In what follows, we will prove that the solution of the relaxed problem (4.68) is also the one of the original problem (4.65).

**Theorem 4.13** *All the variables in $\boldsymbol{\Phi}(t)$ are met with 0 or 1 for the solution of problem (4.68).*

***Proof*** We assume that the solution of (4.68) contains a variable $\phi_{i,k}(t)$ with $0 < \phi_{i,k}(t) < 1$. For simplicity, we denote the slope of $\phi_{i,k}(t)$ in the objective function (4.68a) by $X_{i,k} = \log_2\left(1 + \frac{P^k_{i,BS}(t)}{\sigma^2 + I_{k,U2U}(t)}\right)$, where $X_{i,k} > 0, \forall i \in N, k \in K$. When the objective function is maximized, at least one of the constraints between (4.68b) and (4.68c) is met with equality. In the following, we separate the problem into two conditions, and discuss them successively.

1. *Only One Constraint is Met with Equality*: Without loss of generality, we assume that only (4.68b) is met with equality. Since $\phi_{i,k}(t)$ is not an integer, there exists another variable $\phi_{j,k}(t)$ that is also non-integer to meet the constraint equality of (4.68b). We assume that $X_{i,k} > X_{j,k}$. When we increase $\phi_{i,k}(t)$ and decrease $\phi_{j,k}(t)$ within the constraint, the objective function will be improved. Thus, the solution with $0 < \phi_{i,k}(t) < 1$ is not the optimal solution.
2. *Both (4.68b) and (4.68c) are Met With Equality*: When both (4.68b) and (4.68c) are met with equality, there are at least three more variables that are non-integer to meet the constraint equality. We denote the other three variables by $\phi_{j,k}(t)$, $\phi_{i,m}(t)$, and $\phi_{j,m}(t)$. If $X_{i,k} + X_{j,m} > X_{i,m} + X_{j,k}$, when we increase $\phi_{i,k}(t)$ and $\phi_{j,m}(t)$, and decrease $\phi_{j,k}(t)$ and $\phi_{i,m}(t)$, the objective function will be improved. If $X_{i,k} + X_{j,m} < X_{i,m} + X_{j,k}$, the opponent adjustment will improve the objective function. As a result, the current solution is not the optimal one.

In conclusion, the solution that contains $0 < \phi_{i,k}(t) < 1$ is not the optimal one. When the optimal solution of (4.68) is achieved, all the variables in $\boldsymbol{\Phi}(t)$ are either 0 or 1. □

As shown in Theorem 4.13, the solution of problem (4.68) is either 0 or 1, which satisfies the constraint (4.65d) of the original problem. Therefore, the relaxation of variable $\phi_{i,k}(t)$ does not affect the solution of the subproblem (4.65). The solution of the relaxed problem (4.68) with CVX method is equivalent to the solution of problem (4.65).

#### 4.3.4.2 U2U Subchannel Allocation Algorithm

In this subsubsection, we focus on solving the U2U subchannel allocation subproblem (4.66). We first substitute (4.50), (4.51), (4.52), (4.56), and (4.57) into (4.66a), and the objective function is given by

$$\begin{aligned}\max_{\boldsymbol{\Psi}(t)} \sum_{k=1}^{K} \Bigg( \sum_{\substack{i=1\\ i\in\lambda(t)}}^{N_h} \phi_{i,k}(t) \log_2 \left(1 + \frac{P_{i,BS}^k(t)}{\sigma^2 + \sum_{\substack{j=1\\ j\in\lambda(t)}}^{N_l} \psi_{j,k}(t) P_{j,BS}^k(t)}\right) \\ + \sum_{i=N_h+1}^{N_h+M} \phi_{i,k}(t) \log_2 \left(1 + \frac{P_{i,C}^k(t)}{\sigma^2 + \sum_{\substack{j=1\\ j\in\lambda(t)}}^{N_l} \psi_{j,k}(t) P_{j,BS}^k(t)}\right)\Bigg).\end{aligned} \tag{4.69}$$

When substituting (4.62) and (4.63) into constraint (4.66b), it can be expanded as

$$R_{U2U,i} = \sum_{k=1}^{K} \psi_{i,k}(t) \log_2 \left(1 + \frac{P_U G(d_{i,j}(t))^{-\alpha}}{A+B}\right) \geq R_0, \forall i,\ j \in N, \xi_{i,j} = 1, \tag{4.70}$$

where $A = \sigma^2 + \sum_{n=1,n\in\lambda(t)}^{N_h} \phi_{n,k}(t) I_{m,UAV}^k(t) + \sum_{m=N_h+1}^{N_h+M} \phi_{m,k}(t) I_{m,C}^k(t)$ is fixed in this subproblem, and $B = \sum_{m=1,m\neq i}^{N_l} \psi_{m,k}(t) P_U G(d_{m,j}(t))^{-\alpha}$. Problem (4.66) is a *0-1* programming problem, which has been proved to be NP-hard [38]. In addition, due to the interference from different U2U links, the continuity relaxed problem of (4.66) is still non-convex with respect to $\boldsymbol{\Psi}(t)$. Therefore, problem (4.66) cannot be solved by the existing convex techniques. To solve problem (4.66) efficiently, we utilize the branch-and-bound method [39].

To facilitate understanding of the branch-and-bound algorithm, we first introduce important concepts of fixed and unfixed variables.

**Definition 4.1** When the value of a variable that corresponds to the optimal solution is ensured, we define it as a **fixed variable**. Otherwise, it is an **unfixed variable**.

The solution space of U2U subchannel pairing matrix $\boldsymbol{\Psi}(t)$ can be considered as a binary tree. Each node of the binary tree contains the information of all the variables in $\boldsymbol{\Psi}(t)$. At the root node, all the variables in $\boldsymbol{\Psi}(t)$ are unfixed. The value of an unfixed variable at a father node can be either 0 or 1, which branches the node into two child nodes. Our objective is to search the binary tree for the optimal solution of problem (4.66). The key idea of the branch-and-bound method is to prune the infeasible branches and approach the optimal solution efficiently.

At the beginning of the algorithm, we obtain a feasible solution of problem (4.66) by a proposed low-complexity feasible solution searching (LFSS) method, and set it as the lower bound of the solution. We then start to search the optimal solution of problem (4.66) in the binary tree from its root node. On each node, the branch-and-bound method consists two steps: bound calculation and variable fixation. In the bound calculation step, we evaluate the upper bound of the objective function and the bounds of the constraints separately to prune the branches that cannot achieve a feasible solution above the lower bound of the solution. In the variable fixation step, we fix the variables which has only one feasible value that satisfies the bound requirements in the bound calculation step. We then search the node that contains the newly fixed variables, and continue the two steps of bound calculation and variable fixation. The algorithm terminates when we obtain a node with all the variables fixed. In what follows, we first introduce the LFSS method to achieve the initial feasible solution, and then describe the bound calculation and variable fixation process at each node in detail. Finally, we summarize the branch-and-bound method.

1. *Initial Feasible Solution Search*: In what follows, we propose the LFSS method to obtain a feasible solution of problem (4.66) efficiently. Each UAV in the U2U mode requests a subchannel until its minimum U2U rate threshold is satisfied, and the BS assigns the requested subchannel to the corresponding UAV in the LFSS. The detailed description is shown in Algorithm 8.

Given the U2N and CU subchannel assignment, each UAV in the U2U mode can make a list of data rate that it may achieve from every subchannel without considering the potential U2U interference. The UAVs then sort the subchannels in descending order of achievable rate. We then calculate the data rate of each U2U link with U2N, CU, and U2U interference when the UAVs are assigned to their most preferred subchannels. If the data rate of an UAV is still below the minimum threshold, the UAV will be assigned to its most preferred subchannel which has not been paired with. The subchannel assignment ends when the minimum U2U rate threshold (4.66c) is satisfied by every UAV in the U2U mode. Finally, we adopt the current U2U subchannel pairing result as the initial feasible solution.

2. *Bound Calculation*: In this part, we describe the process of bound calculation at each node. After the initialization step, we start bound calculation from the root node, in which all the variables in $\boldsymbol{\Psi}(t)$ are unfixed, i.e., the value of each $\psi_{i,k}(t)$ in the optimal solution is unknown. We first define a branch pruning operation which is performed in the following bound calculation step.

**Definition 4.2** When a node is **fathomed**, all its child nodes cannot be the optimal solution of the problem.

---

**Algorithm 8:** Initial feasible solution for U2U subchannel allocation

---

**begin**
  Each UAV in the U2U mode calculates its data rate over every subchannel with U2N and CU interference;
  Each UAV sorts the subchannels in descending order of achievable rate;
  Assign the UAVs with their most preferred subchannel;
  Calculate the data rate of each U2U link with U2N, CU, and U2U interference;
  **while** *The data rate of an UAV does not satisfy U2U rate constraint (4.66c)* **do**
    Assign the UAV to its most preferred subchannel that has not been paired;
  **end**
  Set the current U2U-subchannel pairing result as the initial feasible solution;
**end**

---

We calculate the bounds of the objective function and the constraints separately. For simplicity, we denote the objective function with U2U subchannel matrix by $f(\boldsymbol{\Psi}(t))$, and the lower bound of the solution by $f^{lb}$. In what follows, we will elaborate the detailed steps of the bound calculation at each node.

**Step 1 Objective Bound Calculation** The upper bound of the objective function (4.66a) is given as

$$
\begin{aligned}
\bar{f} = & \sum_{k=1}^{K} \Bigg( \sum_{\substack{i=1 \\ i \in \lambda(t)}}^{N_h} \phi_{i,k}(t) \log_2 \left( 1 + \frac{P_{i,BS}^{k}(t)}{\sigma^2 + \sum_{j=1}^{N_l} \psi_{j,k}^{F}(t) P_{j,BS}^{k}(t)} \right) \\
& + \sum_{i=N_h+1}^{N_h+M} \phi_{i,k}(t) \log_2 \left( 1 + \frac{P_{i,C}^{k}(t)}{\sigma^2 + \sum_{j=1}^{N_l} \psi_{j,k}^{F}(t) P_{j,BS}^{k}(t)} \right) \Bigg),
\end{aligned} \tag{4.71}
$$

where $\psi_{j,k}^{F}(t)$ is the fixed variables in the current node, i.e., we ignore the U2U interference of the unfixed variables. If the upper bound of the current node is below the lower bound of the solution, i.e., $\bar{f} < f^{lb}$, we fathom the current node and backtrack to an unfathomed node with unfixed variable. If the current node is not fathomed by the objective function bound calculation, we move to step 2 to check the bounds of the constraints.

**Step 2 Constraint Bounds Calculation** For each UAV in the U2U mode, the upper bound of its U2U rate needs to be larger than the minimum U2U rate threshold. The upper bound of U2U rate for UAV $i$ is achieved when we set all the unfixed variables of UAV $i$ as 1, and all the unfixed variables of other UAVs as 0, which can be expressed as

$$
\bar{R}_{U2U,i} = \sum_{k=1}^{K} \psi_{i,k}(t)|_{\{\psi_{i,k}^{U}(t)=1\}} \log_2 \left( 1 + \frac{P_U G(d_{i,j}(t))^{-\alpha}}{A + B^F} \right), \forall i, j \in N, \xi_{i,j} = 1, \tag{4.72}
$$

where $\psi^{U}_{m,k}(t)$ is the unfixed variables in the current node, and

$$B^F = \sum\nolimits_{m=1, m \neq i}^{N_l} \psi^{F}_{m,k}(t) P_U G(d_{m,j}(t))^{-\alpha}. \tag{4.73}$$

If $\exists \xi_{i,j} = 1$, $\bar{R}_{U2U,i} < R_0$, the minimum U2U rate threshold cannot be satisfied, and the current node is fathomed. Moreover, if there exists a UAV $i$ that does not satisfy constraint (4.66c), i.e., $\sum_{k=1}^{K} \psi_{i,k}(t) > \chi_{max}, \forall i \in N$, the current node is also fathomed. We then backtrack to an unfathomed node in the binary tree and perform bound calculation at the new node.

In the bound calculation procedure, if the objective function of a U2U subchannel pairing matrix $f(\tilde{\boldsymbol{\Psi}}(t))$ is found to be larger than the lower bound of the solution $f^{lb}$, and $\tilde{\boldsymbol{\Psi}}(t)$ satisfies all the constraints, we replace the lower bound of the solution with $f^{lb} = f(\tilde{\boldsymbol{\Psi}}(t))$ to improve the algorithm efficiency. A higher lower bound of the solution helps us to prune the infeasible branches more efficiently.

3. *Variable Fixation*: For a node that is not fathomed in the bound calculation steps, we try to prune the branches by fixing the unfixed variables as follows. The variable fixation is completed in two steps, namely objective fixation and U2U constraint fixation.

**Step 1 Objective Fixation** In the objective fixation process, we denote the reduction of the upper bound when fixing a free variable $\psi_{i,k}(t)$ at 0 or 1 by $p^0_{i,k}$ and $p^1_{i,k}$, respectively. For each unfixed variable $\psi_{i,k}(t)$, we compute $p^0_{i,k}$ and $p^1_{i,k}$ associated with the upper bound $\bar{f}$. If $\bar{f} - p^0_{i,k} \leq f_{opt}$, it means that when we set $\psi_{i,k}(t) = 0$, the upper bound of the child node will fall below the temporary feasible solution. Therefore, we prune the branch of $\psi_{i,k}(t) = 0$, and fix $\psi_{i,k}(t) = 1$. Similarly, if $\bar{f} - p^1_{i,k} \leq f_{opt}$, we prune the branch of $\psi_{i,k}(t) = 1$, and fix $\psi_{i,k}(t) = 0$.

**Step 2 U2U Constraint Fixation** In the U2U constraint fixation process, we denote the U2U rate upper bound reduction for UAV $i$ when fixing a free variable $\psi_{i,k}(t)$ at 0 by $q^0_{i,k}$. If inequality $\bar{R}_{U2U,i} - q^0_{i,k} < R_0$ is satisfied, it means that only when subchannel $k$ is assigned to UAV $i$, the minimum U2U rate threshold of UAV $i$ is possible to be satisfied. Therefore, we prune the branch of $\psi_{i,k}(t) = 0$ and fix $\psi_{i,k}(t) = 1$.

In the objective fixation step and the U2U constraint fixation step, variable $\psi_{i,k}(t)$ may be fixed at different values, which implies that neither of the two child nodes satisfy the objective bound relation and the constraint bound relation simultaneously. Therefore, we fathom the current node and backtrack to an unfathomed node with unfixed variable.

After performing the variable fixation step of the current node, if at least one unfixed variable is fixed at a certain value in the above procedure, we move to the corresponding child node, and continue the algorithm by performing bound calculation and variable fixation at the new node. Otherwise, we generate two new nodes by setting an unfixed variable at $\psi_{i,k}(t) = 0$ and $\psi_{i,k}(t) = 1$, respectively.

**Algorithm 9:** Branch-and-bound method for U2U subchannel allocation

**Input**: The U2N subchannel allocation matrix $\boldsymbol{\Phi}(t)$; The UAV trajectories $\boldsymbol{\omega}(t)$;
**Output**: The U2U subchannel allocation matrix $\boldsymbol{\Psi}(t)$;

**begin**
  **Initialization:** Compute an initial feasible solution $\boldsymbol{\Psi}(t)$ to problem (4.66) and set it as the lower bound of the solution;
  Perform bound calculation and variable fixation at the root node;
  **while** *Not all variables have been fixed* **do**
    Bound calculation;
    **if** *The bound constraints cannot be satisfied* **then**
      Fathom the current node and backtrack to an unfathomed node with unfixed variable;
    **end**
    Variable fixation;
    **if** *At least one variable can be fixed* **then**
      **Go to** the node with newly fixed variable;
    **else**
      Generate two new nodes by setting an unfixed variable $\psi_{i,k}(t) = 0$ and $\psi_{i,k}(t) = 1$;
      **Go to** one of the two nodes firstly;
    **end**
  **end**
  The fixed variables are the final output of $\boldsymbol{\Psi}(t)$;
**end**

We then move to one of the two nodes and continue the algorithm. The branch-and-bound algorithm is accomplished when all variables have been fixed, and the fixed variables are the final solution.

The branch-and-bound method that solves the U2U subchannel allocation subproblem (4.66) is summarized in Algorithm 9.

#### 4.3.4.3 UAV Speed Optimization Algorithm

In the following, we will introduce how to solve the UAV speed optimization subproblem (4.67). The problem is difficult to be optimized directly due to the complicated expression of the air-to-ground transmission model and the change of interference caused by the move of the UAVs. In the following, we first raise two rational assumptions that simplifies this problem, and then propose an efficient solution that gives an approximate solution for problem (4.67).

1. *Two Basic Assumptions*: In this part, we give two assumptions to simplify the UAV speed optimization problem.

We first assume that the pathloss variables $PL_{LoS,i}(t)$ and $PL_{NLoS,i}(t)$ changes much faster than the LoS probability variables $P_{LoS,i}(t)$ and $P_{NLoS,i}(t)$ with the move of a UAV.

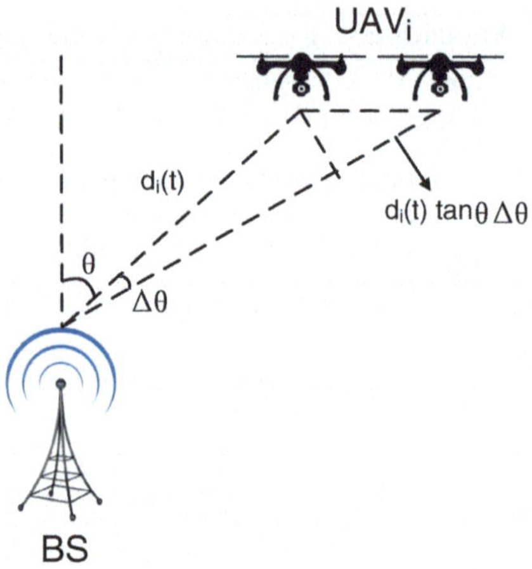


**Fig. 4.18** Transmission model variation with UAV movement

***Proof*** As shown in Fig. 4.18, when the elevation angle of a UAV changes for $\Delta\theta$, e.g., from $\theta$ to $\theta + \Delta\theta$, with $\theta \gg \Delta\theta$, the change of the transmission distance can be approximated as $d_i(t)\tan\theta\,\Delta\theta$. According to Eq. (4.47), the rate of change of the LoS probability to the elevation angle is given as

$$\frac{\Delta P_{LoS,i}(t)}{\Delta\theta} = \frac{ab\exp(-b(\theta - a))}{(1 + a\exp(-b(\theta - a)))^2}. \tag{4.74}$$

The relation between the pathloss and the transmission distance is shown in (4.45), and the rate of change of the pathloss to the elevation angle is

$$\begin{aligned}\frac{\Delta PL_{LoS,i}(t)}{\Delta\theta} &= \frac{20\log(d_i(t)\tan\theta\,\Delta\theta) - 20\log(d_i(t))}{\Delta\theta} \\ &= \frac{20\log(\tan\theta\,\Delta\theta)}{\Delta\theta}.\end{aligned} \tag{4.75}$$

When substituting the typical value of $a$, $b$, and $\theta$ into (4.74) and (4.75), we have $\frac{\Delta P_{LoS,i}(t)}{\Delta\theta} \ll \frac{\Delta PL_{LoS,i}(t)}{\Delta\theta}$. Therefore, the channel pathloss varies much faster than the LoS probability with the movement of a UAV. □

Moreover, we assume that the U2U transmission distance is much larger than the moving distance of a UAV in one time slot, i.e., $d_{i,j}(t) \gg v_{max}$.

With the above two assumptions, we then introduce the solution that solves problem (4.67) efficiently. Note that in problem (4.67), the speed optimization of a pair of U2U transmitting and receiving UAVs is related to constraint (4.67b), but the speed optimization of the UAVs in the U2N mode that do not receive U2U transmissions is irrelevant to constraint (4.67b). Therefore, the speed optimization of the UAVs can be separated into two types: non-U2U participated UAVs and U2U participated UAVs that contains the transmitting UAVs and the corresponding receiving UAVs.

2. *Non-U2U Participated UAV Speed Optimization*: For non-U2U participated UAVs, constraint (4.67b) is not considered. We denote the length of trajectory that UAV $i$ has flied before time slot $t$ by $\mathscr{L}_i(t)$. To satisfy constraint (4.67c) and (4.67d), the length of trajectory that UAV $i$ needs to move along in the following time slots should be no more than the number of following time slots $T-t-1$ times the maximum UAV speed $v_{max}$, i.e., $L_i - \mathscr{L}_i(t+1) < v_{max} \times (T-t-1)$. Therefore, the feasible range of UAV $i$'s speed in time slot $t$ is $\min\{0, L_i - \mathscr{L}_i(t) - v_{max} \times (T-t-1)\} \leq v_i(t) \leq v_{max}$. Problem (4.67) can be simplified as

$$\max_{v_i(t)} \sum_{k=1}^{K} \sum_{\substack{i=1 \\ i\in\lambda(t)}}^{N_h+M} \phi_{i,k}(t) R_{i,BS}^{k}(t), \tag{4.76a}$$

$$\min\{0, L_i - \mathscr{L}_i(t) - v_{max} \times (T-t-1)\} \leq v_i(t) \leq v_{max}. \tag{4.76b}$$

With these two basic assumptions, we can assume that the probability of the LoS and NLoS connections do not change prominently in a single time slot, and the uplink rate is determined by the LoS and NLoS pathloss given in (4.45) and (4.46). Therefore, problem (4.76) is approximated as a convex problem, and can be solved with existing convex optimization methods.

3. *U2U Participated UAV Speed Optimization*: In this part, we introduce the speed optimization of a pair of UAVs: UAV $i$ and UAV $j$, with $\xi_{i,j} = 1$. In time slot $t$, UAV $i$ performs U2U transmission and send the collected data to UAV $j$. UAV $j$ receives the data from UAV $i$, and performs U2N transmission simultaneously. Similarly, constraint (4.67c) and (4.67d) can be simplified as $\min\{0, L_i - \mathscr{L}_i(t) - v_{max} \times (T-t-1)\} \leq v_i(t) \leq v_{max}$, and $\min\{0, L_j - \mathscr{L}_j(t) - v_{max} \times (T-t-1)\} \leq v_j(t) \leq v_{max}$ for UAV $i$ and UAV $j$, respectively. Given the subchannel pairing matrices $\boldsymbol{\Phi}(t)$ and $\boldsymbol{\Psi}(t)$, the U2U rate constraint (4.67b) can be transformed to a distance constraint. When substituting (4.62) and (4.63) into (4.67b), the U2U rate constraint can be shown as

$$d_{i,j}(t) \leq \frac{P_U G}{\left(\sigma^2 + \sum_{m=1,m\neq i}^{N_l+N_h} I_{m,UAV}^{k}(t) + \sum_{m=1}^{M} I_{m,C}^{k}(t)\right)} \frac{1}{\left(2^{\frac{R_0}{\sum_{k=1}^{K} \psi_{i,k}(t)}} - 1\right)}. \tag{4.77}$$

Given the second basic assumption, the U2U interference can be approximated to a constant in each single time slot. Therefore, the right side of Eq. (4.77) can be regarded as a constant, denoted by $d_{i,j}^{max}$ for simplicity. Given the feasible speed range of UAV $i$ and the maximum distance between UAV $i$ and UAV $j$, i.e., $d_{i,j}^{max}$, a feasible speed range of UAV $j$ in time slot $t$ can be obtained, which is written as $v_j(t)^{min} \leq v_j(t) \leq v_j(t)^{max}$. The UAV speed optimization subproblem is reformulated as

$$\max_{v_i(t)} \sum_{k=1}^{K} \sum_{\substack{i=1 \\ i\in\lambda(t)}}^{N_h+M} \phi_{i,k}(t) R_{i,BS}^k(t), \tag{4.78a}$$

$$\min\{0, L_i - \mathscr{L}_i(t) - v_{max} \times (T-t-1)\} \leq v_i(t) \leq v_{max}, \tag{4.78b}$$

$$\min\{0, L_j - \mathscr{L}_j(t) - v_{max} \times (T-t-1)\} \leq v_j(t) \leq v_{max}, \tag{4.78c}$$

$$v_j(t)^{min} \leq v_j(t) \leq v_j(t)^{max}. \tag{4.78d}$$

Similar with problem (4.76), problem (4.78) can also be considered as a convex problem, which can be solved with the existing convex optimization methods.

#### 4.3.4.4 Iterative Subchannel Allocation and UAV Speed Optimization Algorithm

In this subsubsection, we introduce the ISASOA to solve problem (4.64), where U2N and CU subchannel allocation, U2U subchannel allocation, and UAV speed optimization subproblems are solved iteratively. In time slot $t$, we denote the optimization objective function after the $r$th iteration by $\mathscr{R}\Big(\boldsymbol{\Phi}^r(t), \boldsymbol{\Psi}^r(t), \boldsymbol{v}^r(t)\Big)$. In iteration $r$, the U2N and CU subchannel allocation matrix $\boldsymbol{\Phi}(t)$, the U2U subchannel allocation matrix $\boldsymbol{\Psi}(t)$, and the UAV speed variable of UAV $i$ are denoted by $\boldsymbol{\Phi}^r(t)$, $\boldsymbol{\Psi}^r(t)$, and $v_i^r(t)$, respectively. The process of the iterative algorithm for each single time slot is summarized in Algorithm 10.

In time slot $t$, we firstly set the initial condition, where all the subchannels are vacant, and the speed of all the UAVs are given as a fixed value $v_0$, i.e., $\boldsymbol{\Phi}^0(t) = \{0\}$, $\boldsymbol{\Psi}^0(t) = \{0\}$, and $v_i^0(t) = \{v_0\}$, $\forall i \in \mathscr{N}$. We then perform iterations of subchannel allocation and UAV speed optimization until the objective function converges. In

**Algorithm 10:** Iterative subchannel allocation and UAV speed optimization algorithm

```
1  begin
2      Initialization: Set r = 0, Φ^0(t) = {0}, Ψ^0(t) = {0}, ω_i^0(t) = {0}, ∀i ∈ I(t);
3      while R(Φ^r(t), Ψ^r(t), ω^r(t)) − R(Φ^{r−1}(t), Ψ^{r−1}(t), ω^{r−1}(t)) > ε do
4          r = r + 1;
5          Solve U2N and CU subchannel allocation subproblem (4.65), given Ψ^{r−1}(t) and
           v^{r−1}(t);
6          Solve U2U subchannel allocation subproblem (4.66), given Φ^r(t) and v^{r−1}(t);
7          Solve UAV speed optimization subproblem (4.67), given Φ^r(t) and Ψ^r(t);
8      end
9      Output: Φ^r(t), Ψ^r(t), v^r(t);
10 end
```

each iteration, the U2N and CU subchannel allocation is performed first with the U2U subchannel pairing and UAV speed results given in the last iteration, and the U2N and CU subchannel pairing variables are updated. Next, the U2U subchannel allocation is performed as shown in Sect. 4.3.4.2, with the UAV speed obtained in the last iteration and the U2N and CU subchannel pairing results. Afterwards, we perform UAV speed optimization as described in Sect. 4.3.4.3, given the subchannel pairing results. When an iteration is completed, we will compare the values of the objective function obtained in the last two iterations. If the difference between the values is less than a pre-set error tolerant threshold $\epsilon$, the algorithm terminates and the results of subchannel pairing and UAV speed optimization are obtained. Otherwise, the ISASOA will continue.

In the following, we will discuss the convergence and complexity of the proposed ISASOA.

**Theorem 4.14** *The proposed ISASOA is convergent.*

***Proof*** In the $(r+1)$th iteration, we first perform U2N and CU subchannel allocation, and the optimal U2N and CU subchannel allocation solution is obtained with the given $\boldsymbol{\Psi}^r(t)$ and $v_i^r(t)$. Therefore, we have

$$\mathscr{R}\Big(\boldsymbol{\Phi}^{r+1}(t), \boldsymbol{\Psi}^r(t), \boldsymbol{v}^r(t)\Big) \geq \mathscr{R}\Big(\boldsymbol{\Phi}^r(t), \boldsymbol{\Psi}^r(t), \boldsymbol{v}^r(t)\Big), \tag{4.79}$$

i.e., the total rate of U2N and CU transmissions does not decrease with the U2N and CU subchannel allocation in the $(r+1)$th iteration. When solving U2U subchannel allocation, we give the optimal solution of $\boldsymbol{\Psi}^{r+1}(t)$ with $\boldsymbol{\Phi}^{r+1}(t)$ and $\boldsymbol{v}^r(t)$. The relation between $\mathscr{R}\Big(\boldsymbol{\Phi}^{r+1}(t), \boldsymbol{\Psi}^{r+1}(t), \boldsymbol{v}^r(t)\Big)$ and $\mathscr{R}\Big(\boldsymbol{\Phi}^{r+1}(t), \boldsymbol{\Psi}^r(t), \boldsymbol{v}^r(t)\Big)$ can then be expressed as

$$\mathscr{R}\Big(\boldsymbol{\Phi}^{r+1}(t), \boldsymbol{\Psi}^{r+1}(t), \boldsymbol{v}^r(t)\Big) \geq \mathscr{R}\Big(\boldsymbol{\Phi}^{r+1}(t), \boldsymbol{\Psi}^r(t), \boldsymbol{v}^r(t)\Big). \tag{4.80}$$

The optimal speed for the UAVs with $\boldsymbol{\Phi}^r(t)$ and $\boldsymbol{\Psi}^r(t)$ are obtained in the UAV speed optimization algorithm, which can be expressed as

$$\mathscr{R}\Big(\boldsymbol{\Phi}^{r+1}(t), \boldsymbol{\Psi}^{r+1}(t), \boldsymbol{v}^{r+1}(t)\Big) \geq \mathscr{R}\Big(\boldsymbol{\Phi}^{r+1}(t), \boldsymbol{\Psi}^{r+1}(t), \boldsymbol{v}^r(t)\Big). \tag{4.81}$$

In the $r+1$th iteration, we have the following inequation:

$$\begin{aligned} &\mathscr{R}\Big(\boldsymbol{\Phi}^{r+1}(t), \boldsymbol{\Psi}^{r+1}(t), \boldsymbol{v}^{r+1}(t)\Big) \geq \mathscr{R}\Big(\boldsymbol{\Phi}^{r+1}(t), \boldsymbol{\Psi}^{r+1}(t), \boldsymbol{v}^r(t)\Big) \\ &\geq \mathscr{R}\Big(\boldsymbol{\Phi}^{r+1}(t), \boldsymbol{\Psi}^r(t), \boldsymbol{v}^r(t)\Big) \geq \mathscr{R}\Big(\boldsymbol{\Phi}^r(t), \boldsymbol{\Psi}^r(t), \boldsymbol{v}^r(t)\Big). \end{aligned} \tag{4.82}$$

As shown in (4.82), the objective function does not decrease in each iteration. It is known that such a network has a capacity bound, and the uplink sum-rate cannot increase unlimitedly. Therefore, the objective function has an upper bound, and will converge to a constant after limited iterations, i.e., the proposed ISASOA is convergent. □

**Theorem 4.15** *The complexity of the proposed ISASOA is* $O((N_h(t)+M)\times 2^{N_l(t)})$.

***Proof*** The complexity of the proposed ISASOA is the number of iterations times the complexity of iteration. As shown in Algorithm 10, the objective function increases for at least $\epsilon$ in each iteration. We denote the average uplink sum-rate of the initial solution by $\bar{R}_0(N_h(t), M)$, and the average uplink sum-rate of the ISASOA by $\bar{R}(N_h(t), M)$. The number of iteration is no more than $(\bar{R}(N_h(t), M) - \bar{R}_0(N_h(t), M))/\epsilon$. In addition, the increment of the uplink sum-rate can be expressed as $(\bar{R}(N_h(t), M) - \bar{R}_0(N_h(t), M)) = (N_h(t) + M)\log_2\left(\frac{1+\bar{\gamma}_I}{1+\bar{\gamma}_0}\right)$, where $\bar{\gamma}_I$ is the average SNR of UAVs in the U2N mode and CUs with ISASOA, and $\bar{\gamma}_0$ is the average SNR of UAVs in the U2N mode and CUs with the initial solution. Therefore, the number of iterations is given as $C \times (N_h(t) + M)$, where $C$ is a constant.

In each iteration, the U2N subchannel allocation is solved directly with convex problem solutions. The U2U subchannel allocation is solved with branch-and-bound method, with the complexity being $O(2^{N_l(t)})$. The speed of different UAVs is optimized with convex optimization methods, with a complexity of $O(N_h(t) + N_l(t))$. Therefore, the complexity of each iteration is $O(2^{N_l(t)})$, and the complexity of the proposed ISASOA is $O((N_h(t) + M) \times 2^{N_l(t)})$. □

### 4.3.5 Simulation Results

In this subsection, we evaluate the performance of the proposed ISASOA. The selection of the simulation parameters is based on the existing works and 3GPP specifications[13, 30]. In this simulation, the location of the UAVs is randomly and uniformly distributed in a 3-dimensional area of 2 km × 2 km × $h_{max}$, where $h_{max}$ is the maximum possible height for the UAVs. To study the impact of UAV height on the performance of this network, we simulate two scenarios with $h_{max}$ being 100 and 200 m, respectively. The direction of the pre-determined trajectory for each UAV is given randomly. All curves are generated with over 1000 instances of the proposed algorithm. The simulation parameters are listed in Table 4.5. We compare the proposed algorithm with a greedy subchannel allocation algorithm as proposed in [40]. In the greedy algorithm scheme, the subchannel allocation is performed based on matching theory, and the UAV speed is the same as the proposed ISASOA. The maximum possible height for the UAVs in the greedy algorithm is set as 200 m.

Figure 4.19 depicts the uplink sum-rate with different number of UAVs in the U2N mode. In the proposed ISASOA, the difference between $T = 50$ and $T = 30$ in terms of the uplink sum-rate is about 7%. It is shown that a larger task completion time $T$ corresponds to a higher uplink sum-rate, because the UAVs have larger degree of freedom on the optimization of their speeds with a looser

**Table 4.5** Simulation parameters

| Parameter | Value |
|---|---|
| Number of subchannels $K$ | 10 |
| Number of UAVs in the U2U mode $N_l$ | 5 |
| Number of UAVs $N$ | 20 |
| Number of CUs $M$ | 5 |
| Transmission power $P_U$ | 23 dBm |
| Noise variance $\sigma^2$ | −96 dBm |
| Center frequency | 1 GHz |
| Power gains factor $G$ | −31.5 dB |
| The maximum number of subchannels used by one user $\chi_{max}$ | 2 |
| Algorithm convergence threshold $\epsilon$ | 0.1 |
| U2N channel parameter $\eta_{LoS}$ | 1 |
| U2N channel parameter $\eta_{NLoS}$ | 20 |
| U2N channel parameter $a$ | 12 |
| U2N channel parameter $b$ | 0.135 |
| U2U pathloss coefficient $\alpha$ | 2 |
| Maximum UAV speed $v_{max}$ | 10 m/time slot |
| Length of trajectory $L_i$ | 300 m |
| Minimum U2U rate $R_0$ | 10 bit/(s×Hz) |
| SNR threshold $\gamma_{th}$ | 10 dB |

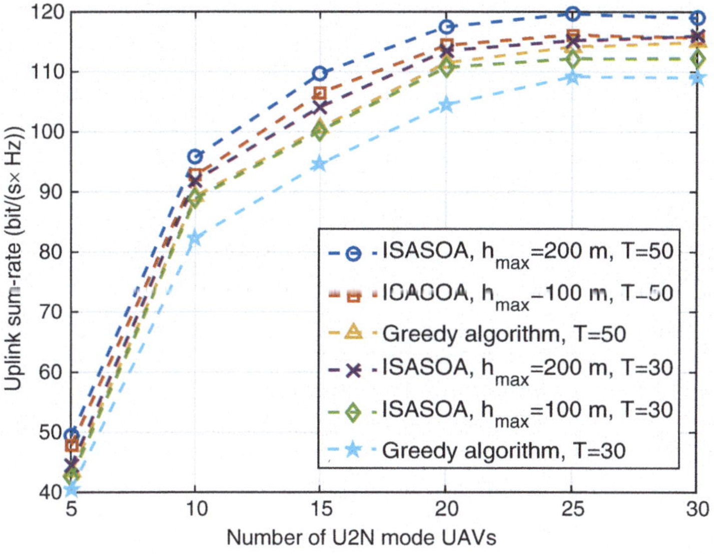


**Fig. 4.19** Number of UAVs in the U2N mode vs. uplink sum-rate

time constraint. The scenario with $h_{max} = 200$ m has about 3% higher uplink sum-rate than the scenario with $h_{max} = 100$ m. The performance gap between the two scenarios is mainly affected by the U2N pathloss caused by different LoS and NLoS probabilities. The uplink sum-rate with the ISASOA is 10% larger than that of the greedy algorithm on average, due to the efficient U2N and U2U subchannel allocation. All the six curves show that the uplink sum-rate of U2N and CU transmissions increases with the number of UAVs, and the growth becomes slower as $N$ increases due to the saturation of network capacity.

Figure 4.20 shows the uplink sum-rate with different U2U-UAV/UAV ratio, when the number of UAVs is set as 20. It is shown that the uplink sum-rate decreases with more UAVs in the U2U mode in the network, and the descent rate is larger with more UAVs in the U2U mode. A larger U2U-UAV/UAV ratio not only reduces the number of UAVs in the U2N mode, but also leads to a larger number of U2U receiving UAVs. Therefore, more UAVs in the U2N mode are restricted by the U2U transmission rate constraint, and cannot move with the speed that corresponds to the maximum rate for the U2N links.

Figure 4.21 illustrates the relation between the U2U-UAV/UAV ratio and the sum-rate for U2U transmissions, with the number of UAVs set at 20. The total U2U transmission rate increases with a larger U2U-UAV/UAV ratio, but the rate of the increment decreases with a larger U2U-UAV/UAV ratio, i.e., the average U2U transmission rate decreases with more UAVs in the U2N mode in the network. The reason is that with the increment of UAVs in the U2N mode, the U2U-to-U2U interference raises rapidly, which reduces the data rate for a U2U link. There is no significant difference between the ISASOA with different $h_{max}$ in terms of

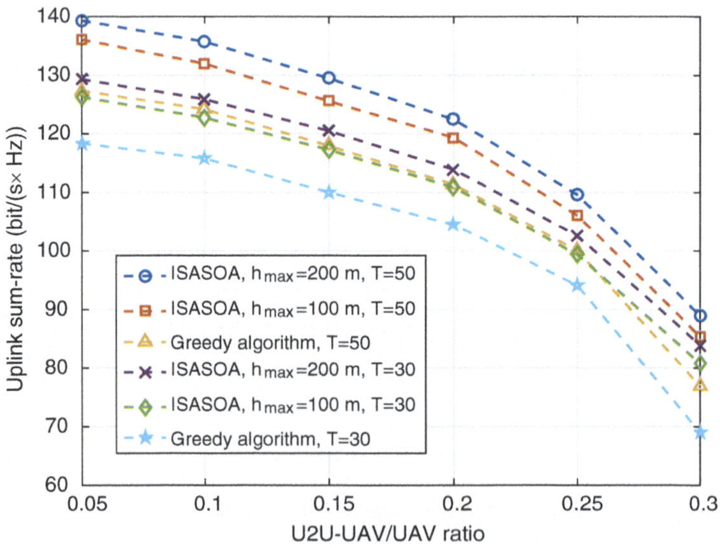


**Fig. 4.20** U2U-UAV/UAV ratio vs. uplink sum-rate

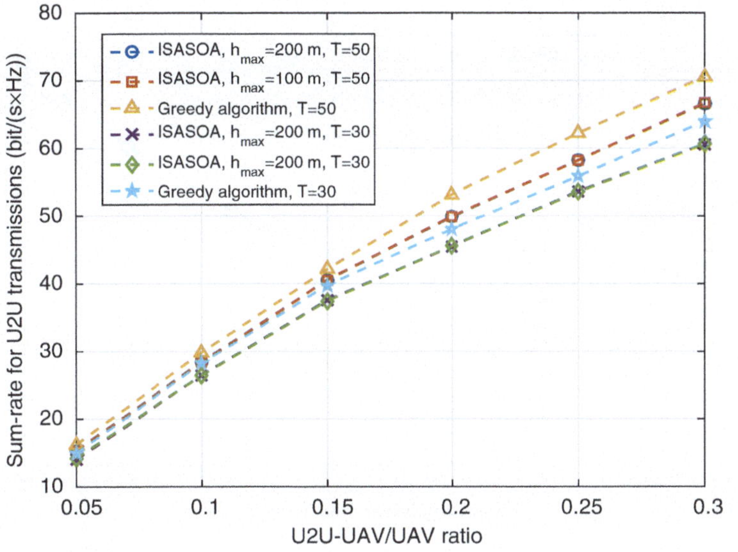


**Fig. 4.21** U2U-UAV/UAV ratio vs. sum-rate for U2U transmissions

the total U2U transmission rate, since the U2U transmission rate for each link is only determined by the distance between the U2U transmitting and receiving UAVs. Note that the average U2U transmission rate is always above the U2U rate threshold within the simulation range. For the greedy algorithm scheme, the total U2U transmission rate is 5% higher than the ISASOA, but a higher U2U transmission rate squeezes the network capacity for the U2N transmissions.

In Fig. 4.22, we give the relation between the task completion time $T$ and the uplink sum-rate. The uplink sum-rate increases with a larger minimum task completion time $T$, and the rate of change increases with $T$. The scheme with $h_{max} = 200$ m has a larger uplink sum-rate than the scheme with $h_{max} = 100$ m due to a higher probability of LoS U2N transmission. The performance gap decreases when $T$ becomes larger, because the UAVs with $h_{max} = 100$ can stay for a longer time at the locations with relatively high LoS transmission probability. The greedy algorithm is about 10% lower than the ISASOA. It can be referred that the uplink sum-rate is affected by the delay tolerance of the data collection.

In Fig. 4.23, the uplink sum-rate is shown with different values of the maximum UAV speed $v_{max}$. A larger maximum UAV speed provides the UAVs a larger degree of freedom on the UAV speed optimization. It is shown that the uplink sum-rate increases significantly with the maximum UAV speed when $v_{max} \leq 20$ m/time slot. The uplink sum-rate turns stable when $v_{max} > 30$ m/time slot, because speed is not the main restriction on the uplink sum-rate when the maximum UAV speed is sufficiently large. The difference between the uplink sum-rate obtained by the ISASOA with $h_{max} = 200$ and that with $h_{max} = 100$ decreases with the increment of $v_{max}$, and the sum-rate obtained by the greedy algorithm is about 10% lower than that obtained by the ISASOA within the simulation range.

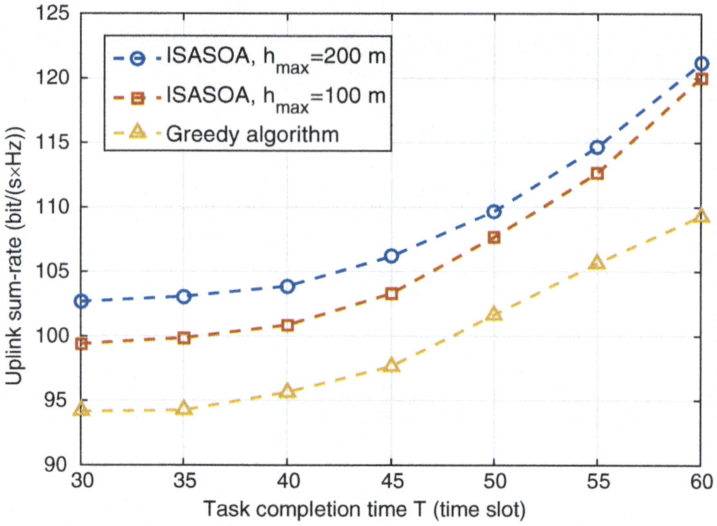


**Fig. 4.22** Minimum task completion time $T$ vs. uplink sum-rate

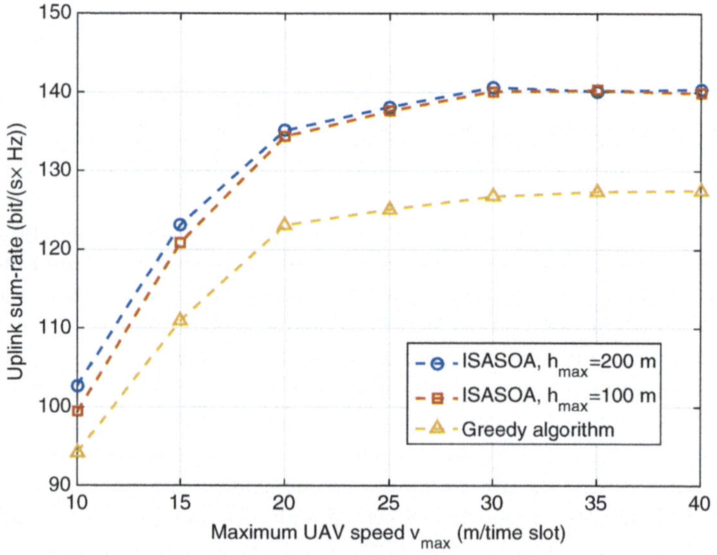


**Fig. 4.23** Maximum UAV speed $v_{max}$ vs. uplink sum-rate

### 4.3.6 Summary

In this section, we have studied a single cell Internet of UAVs, where multiple UAVs upload their collected data to the BS via U2N and U2U transmissions. We have proposed a cooperative UAV sense-and-send protocol and have formulated a joint subchannel allocation and UAV speed optimization problem to improve the uplink sum-rate of the network. To solve the NP-hard problem, we have decoupled it into three subproblems: U2N and CU subchannel allocation, U2U subchannel allocation, and UAV speed optimization. Then, these three subproblems have been solved with optimization methods, and an ISASOA has been proposed to obtain a convergent solution of this problem. Simulation results have shown that the uplink sum-rate decreases with a tighter task completion time constraint, and the proposed ISASOA can achieve about 10% more uplink sum-rate than the greedy algorithm.

## 4.4 Reinforcement Learning for the Cellular Internet of UAVs

In order to enable the real-time sensing applications, the cellular UAV transmission is considered as one promising solution [14, 16], in which the uplink QoS is guaranteed compared to that in ad hoc sensing networks [10]. However, it remains a challenge for UAVs to determine their trajectories independently in such cellular Internet of UAVs. When a UAV is far from the task, it risks in obtaining invalid sensing data, while if it is far from the BS, the low uplink transmission quality may lead to difficulties in transmitting the sensory data back to the BS. Therefore, the UAVs need to take both the sensing accuracy and the uplink transmission quality into consideration in designing their trajectories. Moreover, it is even more challenging when the UAVs belong to different entities and are non-cooperative. Since the spectrum resource is scarce, the UAVs have the incentive to compete for the limited uplink channel resources. Therefore, each UAV should also consider other UAVs that are competing for the spectrum dynamically when it determines the trajectory. Therefore, a decentralized trajectory design approach is necessary for the UAV trajectory design problem, in which the locations of both the task and the BS, as well as the behaviors of the other UAVs should be taken into consideration by each UAV.

To tackle these problems, we consider the scenario where multiple UAVs in a cellular network perform different real-time sensing tasks. We first propose a distributed sense-and-send protocol to coordinate the UAVs, which can be analyzed and by nested Markov chains. Under this condition, the UAV trajectory design problem can be seen as a Markov decision problem, which makes the reinforcement learning the suitable and promising approach to solve the problem. To be specific, we formulate the UAV trajectory design problem under the reinforcement learning framework, and propose an enhanced multi-UAV Q-learning algorithm to solve the problem efficiently.

The rest of this section is organized as follows. In Sect. 4.4.1, the system model is described. In Sect. 4.4.2, we propose a decentralized sense-and-send protocol to coordinate the UAVs performing real-time sensing tasks. We analyze the performance of the proposed sense-and-send protocol in Sect. 4.4.3, and derive the probability of successful valid sensory data transmission using the nested Markov chains. Following that, the reinforcement learning framework and the enhanced multi-UAV Q-learning algorithm are given in Sect. 4.4.4, together with the analyses of complexity, convergence, and scalability. The simulation results are presented in Sect. 4.4.5. Finally, we summarize this section in Sect. 4.4.6.

## 4.4.1 System Model

As illustrated in Fig. 4.24, we consider a single cell orthogonal frequency-division multiple access (OFDMA) cellular Internet of UAVs which consists of $N$ UAVs to perform real-time sensing tasks [41]. We set the horizontal location of the BS as the origin, and the location of the BS and the UAVs can be specified by 3D Cartesian coordinates, i.e., the location of the $i$-th UAV can be denoted as $\boldsymbol{s}_i = (x_i, y_i, h_i)$, and the location of the BS can be denoted as $\boldsymbol{S}_0 = (0, 0, H_0)$ with $H_0$ being its height. Besides, the location of the UAV $i$'s real-time sensing task is denoted as $\boldsymbol{S}_i = (X_i, Y_i, 0)$. To perform the real-time sensing task, each UAV continuously senses the condition of its task, and sends the collected sensory data back to the BS immediately. Therefore, the sensing process and the transmission process jointly determine the UAVs' performance on their real-time sensing tasks. The sensing and transmission models of the UAVs are described in the following.

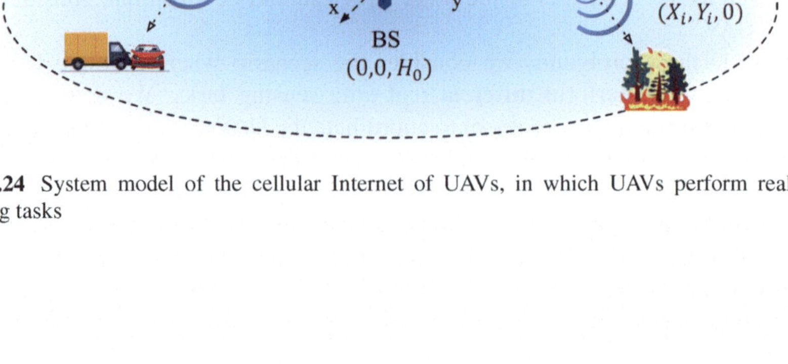


**Fig. 4.24** System model of the cellular Internet of UAVs, in which UAVs perform real-time sensing tasks

#### 4.4.1.1 UAV Sensing

To evaluate the sensing quality of the UAV, we utilize the probabilistic sensing model as introduced in [23, 24], where the successful sensing probability is an exponential function of the distance between the UAV and its task. Supposing that UAV $i$ senses task $i$ for a second, the probability for the UAV to sense the condition of its task successfully, i.e., the successful sensing probability, can be expressed as

$$\mathrm{Pr}_{s,i} = e^{-\lambda l_i}, \tag{4.83}$$

in which $\lambda$ is the parameter evaluating the sensing performance, and $l_i$ denotes the distance from UAV $i$ to task $i$.

It is worth noticing that UAV $i$ cannot figure out whether the sensing is successful or not from its collected sensory data, due to its limited on-board data processing ability. Therefore, UAV $i$ needs to send the sensory data to the BS, and the BS will decide whether the sensory data is valid or not. Nevertheless, UAV $i$ can evaluate its sensing performance by calculating the successful sensing probability based on (4.83).

#### 4.4.1.2 UAV Transmission

In the UAV transmission, the UAVs transmit the sensory data to the BS over orthogonal subchannels to avoid mutual interference. To be specific, we adopt the 3GPP channel model to evaluate the urban macro cellular support for UAVs [13, 42].

Denoting the transmit power of UAVs as $P_u$, we can express the received SNR at the BS of UAV $i$ as

$$\gamma_i = \frac{P_u \| H_i \|}{N_0 10^{\mathrm{PL}_{a,i}/10}}, \tag{4.84}$$

in which $\mathrm{PL}_{a,i}$ denotes the air-to-ground pathloss, $N_0$ denotes the power of noise at the receiver of the BS, and $H_i$ is the small-scale fading coefficient. Specifically, the pathloss $\mathrm{PL}_{a,i}$ and small-scale fading $H_i$ are calculated in two cases separately, i.e., the case where line-of-sight component exists (LoS case), and the case where none LoS components exist (NLoS case). The probability for the UAV $i$-BS channel to contain a LoS component can be calculated as

$$\mathrm{Pr}_{\mathrm{LoS},i} = \begin{cases} 1, & r_i \le r_c, \\ \frac{r_c}{r_i} + e^{-r_i/p_0 + r_c/p_0}, & r_i > r_c, \end{cases} \tag{4.85}$$

in which $r_i = \sqrt{x_i^2 + y_i^2}$, $p_0 = 233.98 \log_{10}(h_i) - 0.95$, and $r_c = \max\{294.05 \log_{10}(h_i) - 432.94, 18\}$.

When the channel contains a LoS component, the pathloss from UAV $i$ to the BS can be calculated as $\text{PL}_{\text{LoS},i} = 30.9 + (22.25 - 0.5\log_{10}(h_i))\log_{10}(d_i) + 20\log_{10}(f_c)$, where $f_c$ is the carrier frequency and $d_i$ is the distance between the BS and UAV $i$. In the LoS case, the small-scale fading $H_i$ obeys Rice distribution with scale parameter $\Omega = 1$ and shape parameter $K[\text{dB}] = 4.217\log_{10}(h_i)+5.787$. On the other hand, when the channel contains none LoS components, the pathloss from UAV $i$ to the BS can be calculated as $\text{PL}_{\text{NLoS},} = 32.4+(43.2-7.6\log_{10}(h_i))\times\log_{10}(d_i) + 20\log_{10}(f_c)$, and the small-scale fading $H_i$ obeys Rayleigh distribution with zero means and unit variance.

To achieve a successful transmission, the SNR at the BS needs to be higher than the decoding threshold $\gamma_{th}$. Therefore, each UAV can evaluate its successful transmission probability by calculating the probability for the SNR at BS to be larger than $\gamma_{th}$. The successful transmission probability $\text{Pr}_{\text{Tx},i}$ for UAV $i$ can be calculated as

$$\text{Pr}_{\text{Tx},i} = \text{Pr}_{\text{Los},i}(1-F_{ri}(\chi_{\text{LoS},i}))+(1-\text{Pr}_{\text{LoS},i})(1-F_{ra}(\chi_{\text{NLoS},i})),$$

in which $\chi_{\text{NLoS},i} = N_0 10^{0.1\text{PL}_{\text{NLoS},i}}\gamma_{th}/P_u$, $\chi_{\text{LoS},i} = N_0 10^{0.1\text{PL}_{\text{LoS},i}}\gamma_{th}/P_u$, $F_{ri}(x) = 1 - Q_1(\sqrt{2K}, x\sqrt{2(K+1)})$ is the CDF of the Rice distribution with $\Omega = 1$ [43], and $F_{ra}(x) = 1 - e^{-x^2/2}$ is the CDF of the Rayleigh distribution with unit variance. Here $Q_1(x)$ denotes the Marcum Q-function of order 1 [44].

### *4.4.2 Decentralized Sense-and-Send Protocol*

In this subsection, we propose a decentralized sense-and-send protocol to coordinate multiple UAVs to perform the sensing tasks simultaneously. We first introduce the sense-and-send cycle, which consists of the beaconing phase, the sensing phase, and the transmission phase. After that, we describe the uplink subchannel allocation mechanism of the BS.

#### 4.4.2.1 Sense-and-Send Cycle

The UAVs perform the sensing tasks in a synchronized iterative manner. Specifically, the process is divided into cycles, which are indexed by $k$. In each cycle, each UAV senses its task and then sends the collected sensory data to the BS. In order to synchronize the transmissions of the UAVs, we further divide each cycle into $T_c$ *frames*, which are the basic time unit for the subchannel allocation. The duration of the time unit frame is set to be the same as that of the transmission and acknowledgement of the sensory data frame.[7]

[7]We assume that the collected sensory data of each UAV in a cycle can be converted into a single sensory data frame with the same length.

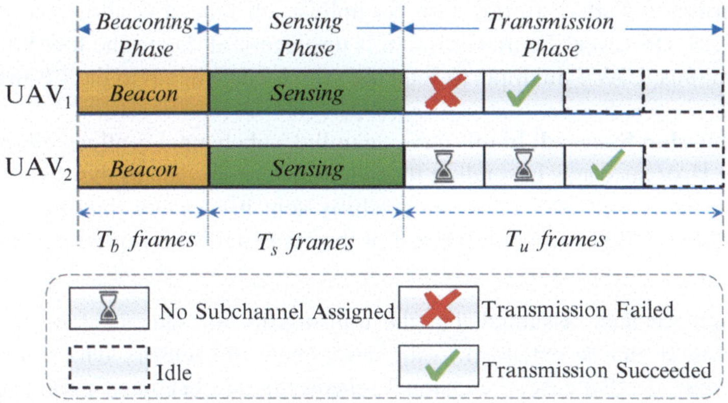


**Fig. 4.25** Decentralized sense-and-send protocol

The cycle consists of three separated phases, i.e., the *beaconing phase*, *sensing phase*, and the *transmission phase*, which contain $T_b$, $T_s$, and $T_u$ frames, respectively. The duration of the beaconing phase and sensing phase is considered to be fixed and determined by the time necessary for transmitting beacons and collecting sensory data. On the other hand, the duration of the transmission phase is decided by the BS, which is related to the network conditions, such as the number of UAVs in the network. Moreover, as illustrated in Fig. 4.25, we consider that the sensing and transmission phases are separated to avoid the possible interference.[8]

In the beaconing phase, each UAV sends its location to the BS in its beacon through the control channel, which can be obtained by the UAV from the on-board GPS positioning. Collecting the beacons sent by the UAVs, the BS then broadcasts to inform the UAVs of the general network settings as well as the locations of all the UAVs. By this means, UAVs can obtain the locations of other UAVs in the beginning of each cycle. Based on the acquired information, each UAV then decides its trajectory in the cycle and informs the BS by another beacon.

In the sensing phase, each UAV senses the task for $T_s$ frames continuously, during which it collects the sensory data. In each frame of the transmission phase, the UAVs attempt to transmit the collected sensory data to the BS if subchannels are allocated to them by the BS. Specifically, there are four possible situations for each UAV, which are shown in Fig. 4.25 and can be described as follows.

- **No Subchannel Allocated**: In this case, no uplink subchannel is allocated to UAV $i$ by the BS. Therefore, the UAV cannot transmit its collected sensory data to the BS. It will wait for the BS to assign a subchannel to it in order to transmit the sensory data.

[8]For example, the UAV's transmission will interfere with its sensing if the UAV tries to sense electromagnetic signals in the frequency bands which are adjacent to its transmission frequency.

- **Transmission Failed**: In this case, an uplink subchannel is allocated to UAV $i$ by the BS. However, the transmission is unsuccessful due to the low SNR at the BS, and thus, UAV $i$ attempts to send the sensory data again to the BS in the next frame.
- **Transmission Succeed**: In this case, an uplink subchannel is allocated to UAV $i$, and the UAV succeeds in sending its collected sensory data to the BS.
- **Idle**: In this case, UAV $i$ has successfully sent its sensory data in the former frames, and will keep idle in the rest of the cycle until the beginning of the next cycle.

Although we have assumed that the transmission of sensory data occupies a single frame, it can be extended to the case where the sensory data transmission takes $n$ frames. In that case, the channel scheduling unit becomes $n$ frames instead of a single frame.

#### 4.4.2.2 Uplink Subchannel Allocation Mechanism

Since the uplink subchannel resources are usually scarce, in each frame of the transmission phase, there may exist more UAVs requesting to transmit the sensory data than the number of available uplink subchannels. To deal with this problem, the BS adopts the following subchannel allocation mechanism to allocate the uplink subchannels to the UAVs.

In each frame, the BS allocates the $C$ available uplink subchannels to the UAVs with uplink requirements, in order to maximize the sum of successful transmission probabilities of the UAVs. Based on the matching algorithm in [45], it is equivalent to that the BS allocates the $C$ available subchannels to the first $C$ UAVs with the highest successful transmission probabilities. The successful transmission probabilities of UAVs can be calculated by the BS based on (4.86), using the information on the trajectories of the UAVs collected in the beaconing phase. Moreover, we denote the transmission states of the UAVs in the $k$-th cycle as the vector $\boldsymbol{I}^{(k)}(t)$, in which $\boldsymbol{I}^{(k)}(t) = (I_1^{(k)}(t), \ldots, I_N^{(k)}(t))$. Here, $I_i^{(k)}(t) = 0$ if UAV $i$ has not succeeded in transmitting its sensory data to the BS at the beginning of the $t$-th frame, otherwise, $I_i^{(k)}(t) = 1$. Based on the above notations, the uplink subchannel allocation can be expressed by the channel allocation vector $\boldsymbol{\nu}^{(k)}(t) = (\nu_1^{(k)}(t), \ldots, \nu_N^{(k)}(t))$, in which the elements are determined by

$$\nu_i^{(k)}(t) = \begin{cases} 1, & Pr_{\text{Tx},i}^{(k)}(t) I_i^{(k)}(t) \geq (\boldsymbol{Pr}_{\text{Tx}}^{k}(t)\boldsymbol{I}^{(k)}(t))_C, \\ 0, & o.w. \end{cases} . \tag{4.86}$$

Here, $\nu_i^{(k)}(t)$ is the *channel allocation indicator* for UAV $i$, i.e., $\nu_i^{(k)}(t) = 1$ only if an uplink subchannel is allocated to UAV $i$ in the $t$-th frame, $Pr_{\text{Tx},i}^{(k)}(t)$ denotes the successful transmission probability of UAV $i$ in the $t$-th frame of the $k$-th cycle, and

$(\boldsymbol{Pr}_{\text{Tx}}^{(k)}(t)\boldsymbol{I}^{(k)}(t))_C$ denotes the $C$-th largest successful transmission probabilities among the UAVs who have not succeeded in sending the sensory data before the $t$-th frame.

Since the trajectory of UAV $i$ determines the distance from UAV $i$ to the BS, it influences the successful transmission probability significantly. As the UAVs are allocated with subchannels only when they have the $C$-largest transmission probabilities, the UAVs have the incentive to compete with each other by selecting trajectories where their successful transmission probabilities are among the highest $C$ ones. Consequently, the UAVs need to design their trajectories with the consideration of not only their distance to the BS and the task, but also of the trajectories of other UAVs.

## 4.4.3 Sense-and-Send Protocol Analysis

In this subsection, we analyze the performance of the proposed sense-and-send protocol by calculating the probability of successful valid sensory data transmission, which plays an important role in solving the UAV trajectory design problem. We first specify the state transitions of the UAVs by using nested bi-level Markov chains. The outer Markov chain depicts the state transitions in the UAV sensing process, and the inner Markov chain depicts the state transitions in the UAV transmission process, which will be elaborated on in the following parts, respectively.

### 4.4.3.1 Outer Markov Chain of UAV Sensing

In the outer Markov chain, the state transition takes place among different cycles. As shown in Fig. 4.26, for each UAV, it has two states in each cycle, i.e., state $\mathscr{H}_f$ to denote that the sensing fails, and state $\mathscr{H}_s$ to denote that the sensing is successful. Supposing the successful sensing probability of UAV $i$ in the $k$-th cycle is $p_{s,i}^{(k)}$, UAV $i$ transits to the state $\mathscr{H}_s$ with probability $p_{s,i}^{(k)}$ and transits to the state

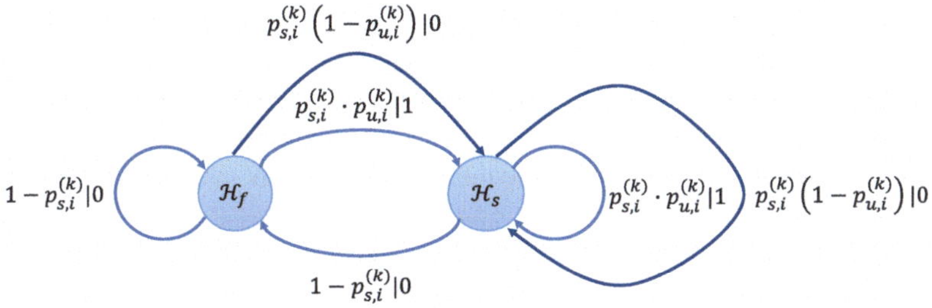


**Fig. 4.26** Illustration on outer Markov chain of UAV sensing

$\mathcal{H}_f$ with probability $(1 - p_{s,i}^{(k)})$ after the $k$-th cycle. The value at the right side of the transition probability denotes the number of valid sensory data that has been transmitted successfully to the BS in the cycle.

Besides, we denote the probability for UAV $i$ to successfully transmit the sensory data to the BS as $p_{u,i}^{(k)}$. Therefore, UAV $i$ successfully transmits valid sensory data to the BS with the probability $p_{s,i}^{(k)} p_{u,i}^{(k)}$. On the other hand, with probability $p_{s,i}^{(k)}(1 - p_{u,i}^{(k)})$, no valid sensory data is transmitted to the BS though the sensing is successful in the $k$-th cycle. The probability $p_{u,i}^{(k)}$ can be analyzed by the inner Markov chain of the UAV transmission in the next subsection, and $p_{s,i}^{(k)}$ can be calculated as follows.

Since the change of the UAVs' locations during each frame is small, we assume that the location of each UAV is fixed within each frame. Therefore, the location of UAV $i$ in the $k$-th cycle can be expressed as a function of the frame index $t$, i.e., $s_i^{(k)}(t) = (x_i^{(k)}(t), y_i^{(k)}(t), h_i^{(k)}(t))$, $t \in [1, T_c]$. Similarly, the distance between UAV $i$ and its task can be expressed as $l_i^{(k)}(t)$, and the distance between the UAV and the BS can be expressed as $d_i^{(k)}(t)$. Moreover, we assume that the UAVs move with a uniform speed and direction in each cycle after the beginning of the sensing phase. Based on the above assumptions, at the $t$-th frame of the $k$-th cycle, the location of UAV $i$ can be expressed as

$$\boldsymbol{s}_i^{(k)}(t) = \boldsymbol{s}_i^{(k)}(T_b) + \frac{t}{T_c}(\boldsymbol{s}_i^{(k+1)}(1) - \boldsymbol{s}_i^{(k)}(T_b)),\ t \in [T_b, T_c]. \tag{4.87}$$

Therefore, the successful sensing probability of UAV $i$ in the cycle can be calculated as

$$p_{s,i}^{(k)} = \prod_{t=T_b+1}^{T_s+T_b} (Pr_{s,i}^{(k)}(t))^{t_f} = \prod_{t=T_b+1}^{T_s+T_b} e^{-\lambda t_f l_i^{(k)}(t)}, \tag{4.88}$$

in which $t_f$ denotes the duration of a frame, and $l_i^{(k)}(t) = \|\boldsymbol{s}_i^{(k)}(t) - \boldsymbol{S}_i\|$.

#### 4.4.3.2 Inner Markov Chain of UAV Transmission

For simplicity of description, we consider a general cycle and omit the superscript cycle index $k$. Since the general state transition diagram is rather complicated, we illustrate the inner Markov chain by giving an example where the number of available uplink subchannel $C = 1$, the number of UAVs $N = 3$, and the number of uplink transmission frames $T_u = 3$.

Taking UAV 1 as an example, the state transition diagram is illustrated in Fig. 4.27. The state of the UAVs in frame $t$ can be represented by the transmission state vector $\boldsymbol{I}(t)$ as defined in Sect. 4.4.2.2. Initially $t = T_b + T_s + 1$, the transmission state is $\boldsymbol{I}(T_b + T_s + 1) = \{0, 0, 0\}$, which indicates that UAVs 1, 2, and 3 have not

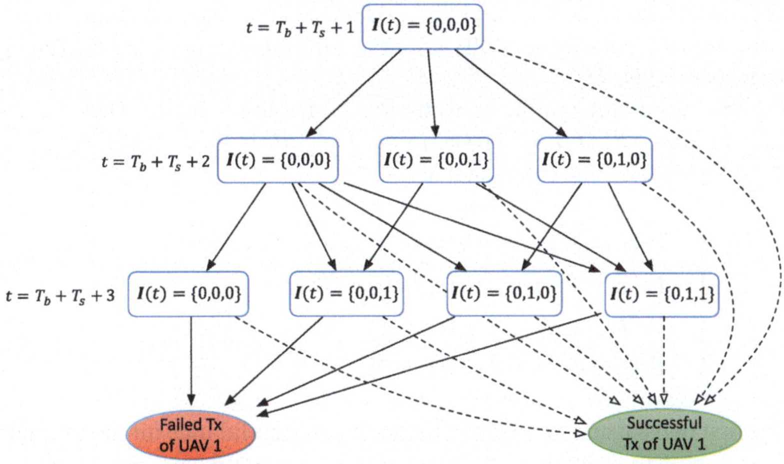


**Fig. 4.27** Illustration on inner Markov chain of UAV 1's transmission given $C = 1$, $N = 3$, $T_u = 3$

succeeded in uplink transmission at the beginning of the transmission phase, and all of them are competing for the uplink subchannel. In the next frame, the transmission state will transit to the *Successful Tx* state for UAV 1, if the sensory data of UAV 1 has been successfully transmitted to the BS. The probability for this transition is equal to $\Pr_{Tx,1}(T_b+T_s+1)\nu_1(T_b+T_s+1)$, i.e., the probability for successful uplink transmission if a subchannel is allocated to UAV 1, otherwise, it is equal to zero.

However, if UAV 1 does not succeed in uplink transmission, the transmission transits into other states, which is determined by whether other UAVs have succeeded in uplink transmission, e.g., it transits to $\boldsymbol{I}(T_b+T_s+2) = (0, 0, 1)$ if UAV 3 succeeds in the first transmission frame. Note that when other UAVs succeed in transmitting sensory data in the previous frames, UAV 1 will face less competitors in the following frames, and thus, it have a larger probability to transmit successfully. Finally, when $t = T_c$, i.e., the last transmission frame in the cycle, UAV 1 will enter the *Failed Tx* state if it does not transmit the sensory data successfully, which means that the sensory data in this cycle is failed to be uploaded. Therefore, to obtain the $p_{u,i}$ in the outer Markov chain is equivalent to calculate the absorbing probability of successful Tx state in the inner Markov chain.

From the above example, it can be observed that the following general recursive equation holds for UAV $i$ when $t \in [T_b + T_s + 1, T_c]$,

$$\begin{aligned}\Pr_{u,i}\{t|\boldsymbol{I}(t)\} = & \Pr_{\mathrm{Tx},i}(t)\nu_i(t)\\ & +\sum_{\substack{\boldsymbol{I}(t+1),\\ I_i(t+1)=0}} \Pr\{\boldsymbol{I}(t+1)|\boldsymbol{I}(t)\}\Pr_{u,i}\{t+1|\boldsymbol{I}(t+1)\},\end{aligned}$$

in which $\Pr\{\boldsymbol{I}(t+1)|\boldsymbol{I}(t)\}$ denotes the probability for the transmission state vector of the $(t+1)$-th frame to be $\boldsymbol{I}(t+1)$, on the condition that the transmission state vector of the $t$-th frame is $\boldsymbol{I}(t)$, and $\Pr_{u,i}\{t|\boldsymbol{I}(t)\}$ denotes the probability for UAV $i$ to transmit sensory data successfully after the $t$-th frame in the current cycle, given the transmission state $\boldsymbol{I}(t)$.

Since the successful uplink transmission probabilities of the UAVs are independent, we have $Pr\{\boldsymbol{I}(t+1)|\boldsymbol{I}(t)\} = \prod_{i=1}^{N} Pr\{I_i(t+1)|I_i(t)\}$, in which $Pr\{I_i(t+1)|I_i(t)\}$ can be calculated as

$$\begin{cases} \Pr\{I_i(t+1)=0|I_i(t)=0\} = 1 - Pr_{\mathrm{Tx},i}(t), \\ \Pr\{I_i(t+1)=1|I_i(t)=0\} = \Pr_{\mathrm{Tx},i}(t), \\ \Pr\{I_t(t+1)=0|I_i(t)=1\} = 0, \\ \Pr\{I_t(t+1)=1|I_i(t)=1\} = 1. \end{cases} \tag{4.89}$$

Here, the first two equations hold due to that the successful transmission probability of UAV $i$ in the $t$-th frame is $\Pr_{\mathrm{Tx},i}(t)$. The third and fourth equations indicate that the UAVs keep idle in the rest of frames once they have successfully sent their sensory data to the BS.

Based on (4.89), the recursive algorithm can be proposed to solve $Pr_{u,i}\{t|\boldsymbol{I}(t)\}$, as presented in Algorithm 11. Therefore, the successful transmission probability can be obtained by $p_{u,i} = Pr_{u,i}\{T_b + T_s + 1|\boldsymbol{I}(T_b + T_s + 1)\}$.

**Algorithm 11:** Algorithm for successful transmission probability in a cycle

**Input**: Frame index ($t$); Transmission state vector ($\boldsymbol{I}(t)$); Length of beaconing phase ($T_b$); Length of sensing phase ($T_s$); Length of transmission phase ($T_u$); Location of UAVs ($\boldsymbol{s}(t)$); Number of subchannels ($C$).
**Output**: $\Pr_{u,i}\{t|\boldsymbol{I}(t)\},\ i = 1, \ldots, N$;

**begin**
  **if** $t = T_b + T_s + 1$ **then**
    $\Pr_{u,i}\{t|\boldsymbol{I}(t)\} := 0, i = 1, \ldots, N$;
  **else**
    **if** $t > T_c$ **then**
      **return** $\Pr_{u,i}\{t|\boldsymbol{I}(t)\} = 0,\ i = 1, \ldots, N$;
    **end**
  **end**
  Calculate the successful transmission probabilities $\Pr_{\mathrm{Tx},i}(t)$, $i = 1, \ldots, N$ based on (4.86);
  Determine the subchannel allocation indicator $\boldsymbol{\nu}(t)$ based on (4.86);
  **for** $i \in [1, N]$ *and* $I_i(t) = 0$ **do**
    $\Pr_{u,i}\{t|\boldsymbol{I}(t)\} := \Pr_{Tx,i}(t)\nu_i(t)$;
  **end**
  **for** *all* $\boldsymbol{I}(t+1)$ *with* $\Pr\{\boldsymbol{I}(t+1)|\boldsymbol{I}(t)\} > 0$ **do**
    Solve $\Pr_{u,i}\{t+1|\boldsymbol{I}(t+1)\}$ by calling Algorithm 11, in which $t := t+1$ and $\boldsymbol{I}(t) := \boldsymbol{I}(t+1)$ and other parameters hold;
    $\Pr_{u,i}\{t|\boldsymbol{I}(t)\} := \Pr_{u,i}\{t|\boldsymbol{I}(t)\} + \Pr\{\boldsymbol{I}(t+1)|\boldsymbol{I}(t)\}\Pr_{u,i}\{t+1|\boldsymbol{I}(t+1)\}$;
  **end**
  **return** $\Pr_{u,i}\{t|\boldsymbol{I}(t)\},\ i = 1, \ldots, N$;
**end**

In summary, the probability of successful valid sensory data transmission for UAV $i$ in the $k$-th cycle can be calculated as

$$p_{\mathrm{sTx},i}^{(k)} = p_{s,i}^{(k)} p_{u,i}^{(k)}. \tag{4.90}$$

#### 4.4.3.3 Analysis on the Data Rate

In this section, we evaluate the data rate by the average number of valid sensory data transmissions per second, which is denoted as $N_{vd}$. The value of $N_{vd}$ is influenced by many factors, such as the distance between the BS and the tasks, the number of available subchannels, the number of UAVs in the network, and the duration of the transmission phase.

Besides, we analyze the influence of the duration of transmission phase $T_u$ on $N_{vd}$ in a simplified case: Assuming all the UAVs are equivalent, i.e., they have the same probabilities for successful uplink transmission in a frame, the same probabilities for successful sensing, and the same probabilities to be allocated subchannels. Based on the above assumptions, the following proposition can be derived.

**Proposition 4.1** *When the $N$ UAVs are equivalent, and have the probability for successful sensing $p_s$, the probability for successful uplink transmission $p_u$, then $N_{vd}$ first increases then decreases with the increment of $T_u$, and the optimal $T_u^*$ can be calculated as*

$$T_u^* = \frac{N}{C\ln(1-p_u)}(1 + W_{-1}(-\frac{(1-p_u)^{\frac{CT_u}{N}}}{e})) - T_b - T_s, \tag{4.91}$$

*in which $W_{-1}(\cdot)$ denotes the lower branch of Lambert-W function in [46].*

***Proof*** Denoting the UAVs' probability for successful uplink transmission as $p_u$ and their probability for successful sensing as $p_s$, we can calculate the average number of valid sensory data transmissions per second by

$$N_{vd} = N \cdot \frac{p_s(1-(1-p_u)^{\frac{CT_u}{N}})}{(T_b + T_s + T_u)t_f},$$

in which $t_f$ is the duration of single frame in seconds.

The partial derivative of $N_{vd}$ with respect to $T_u$ can be calculated as

$$\frac{\partial N_{vd}}{\partial T_u} = \frac{p_s F(T_u)}{t_f(T_b + T_s + T_u)^2}$$

in which $F(T_u) = p_f^{\frac{CT_u}{N}}(N - C(T_b+T_s+T_u)\ln p_f) - N$, and $p_f = 1 - p_u$. Taking partial derivative of $F(T_u)$ with regard to $T_u$, we can derive that $\partial F(T_u)/\partial T_u =$

$-C^2 p_f^{CT_u/N}(T_s + T_b + T_u)\ln p_f/N < 0$. Besides, when $T_u \to \infty$, $F(T_u) \to -N$ and $N_{vd} \to 0$, and when $T_u = 0$, $N_{vd} = 0$. Therefore, $\partial F(T_u)/\partial T_u < 0$ indicates that there is a unique maximum point for $N_{vd}$ when $T_u \in (0, \infty)$.

The maximum of $N_{vd}$ is reached when $F(T_u^*) = 0$, in which $T_u^*$ can be obtained by

$$T_u^* = \frac{N}{C \ln p_f}(1 + W_{-1}(-\frac{p_f^{\frac{CT_u}{N}}}{e})) - T_b - T_s, \tag{4.92}$$

where $W_{-1}(\cdot)$ denotes the lower branch of Lambert-W function [46]. □

The above proposition sheds light on the relation between the spectral efficiency and the duration of transmission phase in the general cases. In the cases where the UAVs are not equivalent, the spectral efficiency also first increases then decreases with the duration of transmission phase. This is because when $T_u = 0$, $N_{vd} = 0$, and when $T_u \to \infty$, $N_{vd} \to 0$.

### 4.4.4 *Decentralized Trajectory Design*

In this section, we first formulate the decentralized trajectory design problem of UAVs, and then analyze the problem in the reinforcement learning framework. After that, we describe the single-agent and multi-agent reinforcement learning algorithms under the framework, and propose an enhanced multi-UAV Q-learning algorithm to solve the UAV trajectory design problem efficiently.

#### 4.4.4.1 UAV Trajectory Design Problem

Before the trajectory design problem formulation, we first set up the model to describe the UAVs' trajectories. In this section, we focus on the cylindrical region with the maximum height $h_{\max}$ and the radius of the cross section $R_{\max}$ which satisfies $R_{\max} = \max\{R_i | R_i = \sqrt{X_i^2 + Y_i^2}, \forall i \in [1, N]\}$, since it is inefficient for the UAVs to move beyond the farthest task. Moreover, we assume that the space is divided into a finite set of discrete spatial points $\mathscr{S}_p$, which is arranged in a square lattice pattern as shown in Fig. 4.28. Therefore, the trajectory of UAV $i$ starting from the $k$-th cycle can be represented as the sequence of spatial points $\mathscr{S}_i^{(k)} = \{\boldsymbol{s}_i^{(k)}, \boldsymbol{s}_i^{(k+1)}, \ldots\}$ that the UAV locates at the beginning of each cycle.

To determine the trajectories, UAVs select their next spatial point at the beginning of each cycle. To be specific, UAV $i$ locates at the spatial point $\boldsymbol{s}_i^{(k)}$ at the beginning of the $k$-th cycle, and decides which spatial point $\boldsymbol{s}_i^{(k+1)}$ it will move to at the beginning of the $(k+1)$-th cycle. After the UAV has selected its next spatial point,

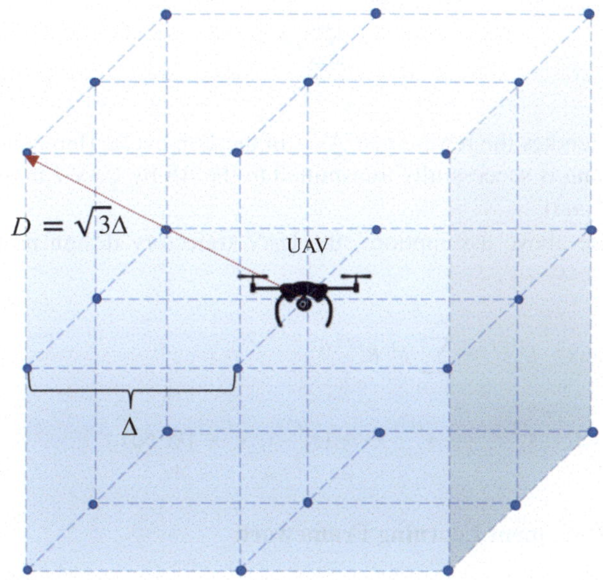


**Fig. 4.28** Illustration on the set of available spatial points that the UAV can reach at the beginning of the next cycle

it will move to the point with a uniform speed and direction within this cycle. Moreover, the available spatial points that UAV $i$ can reach is within the maximum distance it can fly in a cycle, which is denoted as $D$. We set the distance between two adjacent spatial points as $\Delta = D/\sqrt{3}$, and thus, the available spatial point that UAV $i$ can fly to in the $k+1$ cycle is within a cube centered at $(x_i^{(k)}, y_i^{(k)}, h_i^{(k)})$ with side length equal to $2\Delta$, as illustrated in Fig. 4.28.

In Fig. 4.28, there are at most 27 available spatial points that can be selected by the UAVs in each cycle. We denote the set of all the vectors from the center to the available spatial points as the available action set of the UAVs which is denoted as $\mathscr{A}$. Moreover, it is worth noticing that when the UAV is at the marginal location (e.g., flying at the minimum height), there are less available actions to be selected. To handle the differences among the available action sets at different spatial points, we denote the available action set at the spatial point $\boldsymbol{s}$ as $\mathscr{A}(\boldsymbol{s})$.

In this section, we consider the utility of each UAV to be the total number of successful valid sensory data transmissions for its task. Therefore, the UAVs have incentive to maximize the total amount of successful valid sensory data transmission by designing their trajectories. Besides, we assume that the UAVs have discounting valuation on the successfully transmitted valid sensory data. Specifically, for the UAVs in the $k$-th cycle, the successfully valid sensory data transmitted in the $k'$-th cycle is worth only $\rho^{|k'-k|}$ ($\rho \in [0, 1)$) the successful valid sensory data transmitted in the current cycle, due to the timeliness requirements of real-time sensing tasks. Therefore, at the beginning of $k$-th cycle, the utility of UAV $i$ is defined as the *total discounted rewards* in the future, and can be denoted as

$$U_i^{(k)} = \sum_{n=0}^{\infty} \rho^n R_i^{(k+n)}, \tag{4.93}$$

in which $R_i^{(k)}$ denotes the reward of UAV $i$ in the $k$-th cycle. Here, and $R_i^{(k)} = 1$ if valid sensory data is successfully transmitted to the BS by UAV $i$ in the $k$-th cycle, otherwise, $R_i^{(k)} = 0$.

Based on the above assumptions, the UAV trajectory design problem can be formulated as

$$\max_{\mathscr{S}_i^{(k)}} \quad U_i^{(k)} = \sum_{n=0}^{\infty} \rho^n R_i^{(k+n)}, \tag{4.94}$$

$$s.t. \quad \mathbf{s}_i^{(k'+1)} - \mathbf{s}_i^{(k')} \in \mathscr{A}(\mathbf{s}_i^{(k')}),\ k' \in [k, \infty). \tag{4.94a}$$

#### 4.4.4.2 Reinforcement Learning Framework

Generally, the UAV trajectory design problem (4.94) is hard to solve since the rewards of the UAVs in the future cycles are influenced by the trajectories of all UAVs, which are determined in a decentralized manner and hard to model. Fortunately, reinforcement learning is able to deal with the problem of agent programming in environment with deficient understanding, which removes the burden of developing accurate models and solving the optimization with respect to those models [47]. For this reason, we adopt reinforcement learning to solve the UAV trajectory design problem.

To begin with, we introduce a reinforcement learning framework for the problem. With the help of [48], the reinforcement learning framework can be given as follows, in which the superscript cycle index $k$ is omitted for description simplicity.

**Definition 4.3** A reinforcement learning framework for the UAV trajectory design problem is described by a tuple $< \mathscr{S}_1, \ldots, \mathscr{S}_N, \mathscr{A}_1, \ldots, \mathscr{A}_N, \mathscr{T}, p_{R,1}, \ldots, p_{R,N}, \rho >$, where

- $\mathscr{S}_1, \ldots, \mathscr{S}_N$ are finite state spaces of all the possible locations of the UAVs at the beginning of each cycle, and the state space of UAV $i$ is equal to the finite spatial space, i.e., $\mathscr{S}_i = \mathscr{S}_p,\ \forall i \in [1, N]$.
- $\mathscr{A}_1, \ldots, \mathscr{A}_N$ are the corresponding finite sets of actions available to each agents. The set $\mathscr{A}_i$ consists of all the available action of UAV $i$, i.e., $\mathscr{A}_i = \mathscr{A},\ \forall i \in [1, N]$.
- $\mathscr{T} : \prod_{i=1}^N \mathscr{S}_i \times \prod_{i=1}^N \mathscr{A}_i \rightarrow (\mathscr{S}_p)^N$ is the state transition function. It maps the location profile and the action profile of the UAVs in a certain cycle to the location profile of the UAVs in the next cycle.
- $p_{R,i} : \prod_{j=1}^N \mathscr{S}_j \times \prod_{j=1}^N \mathscr{A}_j \rightarrow \Pi(0, 1),\ i = 1, \ldots, N$ represents the reward function for UAV $i$. That is to say, it maps the location profile and action profile

of all the UAVs in the current cycle to the probability for UAV $i$ to get unit reward from successful valid sensory data transmission.

- $\rho \in [0, 1)$ is the discount factor, which indicates UAVs' evaluation of the rewards that obtained in the future or in the past.

In the framework, the rewards of the UAVs are informed by the BS. Specifically, we assume that the BS informs each UAV whether it has transmitted valid sensory data in a certain cycle by the BS beacon at the beginning of the next cycle. For each UAV, it obtains one reward if the BS informs that the valid sensory data has been received successfully. Therefore, the probability for UAV $i$ to obtain one reward is equal to the probability for it to transmit valid sensory data successfully in the cycle, i.e., $p_{R,i} = p_{\text{sTx},i}$. Since the probability of successful valid sensory data transmission is influenced by both the successful sensing probability and the successful transmission probability, the UAV's trajectory learning process is associated with the sensing and transmission processes through the obtained reward in each cycle.

Under the reinforcement learning framework for the UAV trajectory design, the following two kinds of reinforcement learning algorithms can be adopted, which are single-agent Q-learning algorithm and multi-agent Q-learning algorithm.

1. *Single-agent Q-learning Algorithm*: One of the most basic reinforcement learning algorithm is the single-agent Q-learning algorithm [49]. It is a form of model-free reinforcement learning and provides a simple way for the agent to learn how to act optimally. The agent selects actions by following its policy in each state, which is denoted as $\pi(s)$, $s \in \mathscr{S}$ and is a mapping from states to actions. The essence of the algorithm is to find the Q-value of each state and action pairs, which is defined as the accumulated reward received when taking the action in the state and then following the policy thereafter. In its simplest form, the agent maintains a table containing its current estimated Q-values, which is denoted as $Q(s, \boldsymbol{a})$ with $\boldsymbol{a}$ indicating the action. It observes the current state $s$ and selects the action $\boldsymbol{a}$ that maximizes $Q(s, \boldsymbol{a})$ with some exploration strategies. Q-learning has been studied extensively in single-agent tasks where only one agent is acting alone in a stationary environment.

In the UAV trajectory design problem, multiple UAVs take actions at the same time. When each UAV adopts the single-agent Q-learning algorithm, it assumes that the other agents are part of the environment. In the UAV trajectory design problem, the single-agent Q-learning algorithm can be adopted as follows. For UAV $i$, the policy of UAV $i$ to select action is

$$\pi_i(s_i) = \arg \max_{\boldsymbol{a}'_i \in \mathscr{A}(s_i)} Q_i(s_i, \boldsymbol{a}_i). \tag{4.95}$$

Upon receiving a reward $R_i$ after the end of the cycle and observing the next state $s'_i$, it updates its table of Q-values according to the following rule:

**Algorithm 12:** Single-agent Q-learning algorithm for UAV trajectory design problem of UAV $i$

**begin**
  Initialize $Q_i(s_i, a_i) := 0, \forall s_i \in \mathscr{S}_p,\ a_i \in \mathscr{A}_i(s_i), \pi_i(s_i) := \text{Rand}(\{\mathscr{A}_i(s_i)\})$;
  **for** $k = 1$ *to max-number-of-cycles* **do**
    With probability $1 - \epsilon^{(k)}$, choose action $\boldsymbol{a}_i$ from the policy at the state $\pi_i(\boldsymbol{s}_i)$, and with probability $\epsilon^{(k)}$, randomly choose an available action for exploration;
    Perform the action $\boldsymbol{a}_i$ in the $k$-th cycle;
    Observe the transited state $\boldsymbol{s}_i'$ and the reward $R_i$;
    Select $\boldsymbol{a}_i'$ in the transited state $\boldsymbol{s}_i'$ according to $\pi_i(\boldsymbol{s}_i')$;
    Update the Q-value for the former state-action pair, i.e.,
    $Q_i(\boldsymbol{s}_i, \boldsymbol{a}_i) := Q_i(\boldsymbol{s}_i, \boldsymbol{a}_i) + \alpha^{(k)}(R_i + \rho Q(\boldsymbol{s}_i', \boldsymbol{a}_i') - Q_i(\boldsymbol{s}_i, \boldsymbol{a}_i))$;
    Update the policy at state $\boldsymbol{s}_i$ as $\pi_i(\boldsymbol{s}_i, \boldsymbol{a}_i') := 1$, where
    $\boldsymbol{a}_i = \arg\max_{\boldsymbol{m} \in \mathscr{A}_i(\boldsymbol{s}_i)} Q_i(\boldsymbol{s}_i, \boldsymbol{m})$;
    Update the state $\boldsymbol{s}_i := \boldsymbol{s}_i'$;
  **end**
**end**

$$Q_i(\boldsymbol{s}_i, \boldsymbol{a}_i) \leftarrow Q_i(\boldsymbol{s}_i, \boldsymbol{a}_i) + \alpha(R_i + \rho \max_{\boldsymbol{a}_i' \in \mathscr{A}(\boldsymbol{s}_i)} Q_i(\boldsymbol{s}_i, \boldsymbol{a}_i)), \tag{4.96}$$

in which $\alpha \in (0, 1)$ denotes the learning rate. With the help of [50], the single-agent Q-learning algorithm for UAV trajectory design problem can be given in Algorithm 12.

2. *Multi-agent Q-learning Algorithm*: Although single-agent Q-learning algorithm has many favorable properties such as small state space and easy implementation, it lacks consideration on the states and the strategic behaviors of other agents. Therefore, we adopt a multi-agent Q-learning algorithm called opponent modeling Q-learning to solve the UAV trajectory design problem, which enables the agent to adapt to other agents' behaviors.

Opponent modeling Q-learning is an effective multi-agent reinforcement learning algorithm [51, 52], in which explicit models of the other agents are learned as stationary distributions over their actions. These distributions, combined with learned joint state-action values from standard temporal differencing, are used to select an action in each cycle. Specifically, at the beginning of each cycle, UAV $i$ selects the action $a_i$ which maximizes the expected discounted reward according to the observed frequency distribution of other agents' action in the current state $\boldsymbol{s}$, i.e., its policy at $\boldsymbol{s}$ is

$$\pi_i(\boldsymbol{s}) = \arg\max_{\boldsymbol{a}_i'} \sum_{\boldsymbol{a}_{-i}'} \frac{\Phi(\boldsymbol{s}, \boldsymbol{a}_{-i}')}{n(\boldsymbol{s})} Q_i(\boldsymbol{s}, (\boldsymbol{a}_i', \boldsymbol{a}_{-i}')) \tag{4.97}$$

---

**Algorithm 13:** Opponent modeling Q-learning algorithm for UAV trajectory design problem of UAV $i$

---

**begin**
  Initialize $Q_i(\boldsymbol{s}, (\boldsymbol{a}_i, \boldsymbol{a}_{-i})) := 0, \forall \boldsymbol{s} \in \prod_i^N \mathscr{S}_i, \boldsymbol{a}_i \in \mathscr{A}_i(s_i), \boldsymbol{a}_{-i} \in \prod_{j \neq i}^N \mathscr{A}_j$, $\pi_i(s_i) := \text{Rand}(\{\mathscr{A}_i(s_i)\})$;
  **for** $k = 1$ *to max-number-of-cycles* **do**
    With probability $1 - \epsilon^{(k)}$, choose action $\boldsymbol{a}_i$ according to the policy $\pi_i(\boldsymbol{s})$, or with probability $\epsilon^{(k)}$, randomly choose an available action for exploration;
    Perform the action $\boldsymbol{a}_i$ in the $k$-th cycle;
    Observe the transited state $\boldsymbol{s}'$ and the reward $R_i$;
    Select action $\boldsymbol{a}_i'$ in the transited state $\boldsymbol{s}'$ according to the strategy in state $\boldsymbol{s}'$ according to (4.97);
    Update the Q-value for the former state-action pair according to (4.98);
    Update the policy at state $\boldsymbol{s}$ to the action that maximizes the expected discounted reward according to (4.97);
    Update the state $\boldsymbol{s} := \boldsymbol{s}'$;
  **end**
**end**

---

in which the state $\boldsymbol{s} = (s_1, \ldots, s_N)$ is location profile of all the UAVs, $\Phi(\boldsymbol{s}, \boldsymbol{a}'_{-i})$ denotes the number of times for the agents other than agent $i$ to select action profile $\boldsymbol{a}'_{-i}$ in the state $\boldsymbol{s}$, and $n(\boldsymbol{s})$ is the total number of times the state $\boldsymbol{s}$ has been visited.

After the agent $i$ observes the transited state $\boldsymbol{s}'$, the action profile $\boldsymbol{a}_{-i}$, and the reward in the previous cycle, it updates the Q-value as follows:

$$Q_i(\boldsymbol{s}, (\boldsymbol{a}_i, \boldsymbol{a}_{-i})) = (1 - \alpha) Q_i(\boldsymbol{s}, (\boldsymbol{a}_i, \boldsymbol{a}_{-i})) + \alpha (R_i + \rho V_i(\boldsymbol{s}')), \tag{4.98}$$

where $V_i(\boldsymbol{s}') = \max_{\boldsymbol{a}_i'} \sum_{\boldsymbol{a}'_{-i}} \frac{\Phi(\boldsymbol{s}', \boldsymbol{a}'_{-i})}{n(\boldsymbol{s}')} Q(\boldsymbol{s}, (\boldsymbol{a}_i', \boldsymbol{a}'_{-i}))$ indicates that agent $i$ selects the action in the transited state $\boldsymbol{s}'$ to maximize the expected discounted reward based on the empirical action profile distribution. The multi-agent Q-learning algorithm for UAV trajectory design can be given in Algorithm 13.

#### 4.4.4.3 Enhanced Multi-UAV Q-Learning Algorithm for UAV Trajectory Design

In the opponent modeling multi-agent reinforcement learning algorithm, UAVs need to tackle too many state-action pairs, resulting in a slow convergence speed. Therefore, we enhance the opponent modeling Q-learning algorithm in the UAV trajectory design problem by reducing the available action set and adopting a model-based reward representation. These two enhancing approaches are introduced as follows, and the proposed enhanced multi-UAV Q-learning algorithm is given in Algorithm 14.

**Algorithm 14:** Enhanced multi-UAV Q-learning algorithm for trajectory design problem of UAV $i$

**begin**
  **for** $k = 1$ *to max-number-of-cycles* **do**
    Obtain the available action set $\mathscr{A}_j^+(s_j),\ \forall j \in [1, N]$ for the current state $\boldsymbol{s}$ according to Definition 4.4;
    **if** $\boldsymbol{s}$ *has not been visited before* **then**
      Initialize $Q_i(\boldsymbol{s}, \boldsymbol{a}) := p_{\mathrm{sTx},i}(\boldsymbol{s}, \boldsymbol{a}), \forall \boldsymbol{s} \in \prod_i^N \mathscr{S}_i, \boldsymbol{a} \in \prod_{j=1}^N \mathscr{A}_j^+(\boldsymbol{s}_j)$, $\pi_i(\boldsymbol{s}_i) := \mathrm{Rand}(\{\mathscr{A}_i^+(\boldsymbol{s}_i)\})$;
    **end**
    With probability $1 - \epsilon^{(k)}$, choose action $a_i$ from the strategy at the state $\pi_i(\boldsymbol{s})$, or with probability $\epsilon^{(k)}$, randomly choose an available action for exploration;
    Perform action $\boldsymbol{a}_i$ in the $k$-th cycle;
    Observe the transited state $\boldsymbol{s}'$ and the action profile $\boldsymbol{a}$ in the previous state;
    Select action $\boldsymbol{a}_i'$ in the transited state $\boldsymbol{s}'$ according to policy $\pi_i(\boldsymbol{s}')$ defined in (4.97);
    Calculate the probability of successful valid sensory data transmission in the previous cycle $p_{\mathrm{sTx},i}(\boldsymbol{s}, \boldsymbol{a})$;
    Update the Q-function for the former state-action pair according to (4.98), with $R_i := p_{\mathrm{sTx},i}(\boldsymbol{s}, \boldsymbol{a})$;
    Update the policy at state $\boldsymbol{s}$ to the action that maximizes the expected discounted rewards according to (4.97);
    Update the state $\boldsymbol{s} := \boldsymbol{s}'$;
  **end**
**end**

1. *Available Action Set Reduction*: It can be observed that although the UAVs are possible to reach all the spatial points in the finite location space $\mathscr{S}_p$, it makes no sense for the UAVs to move away from the vertical planes passing the BS and their tasks, i.e., the BS-task planes, which decreases the successful sensing probability as well as the successful transmitting probability. Therefore, we confine the available action set of the UAV to the actions which does not increase the horizontal distance between it and the BS-task plane, which is illustrated in Fig. 4.29 (the arrows).

Ideally, the UAVs should be in the BS-task plane and only move within the plane. However, since the spatial space is discrete, the UAV cannot only move within the BS-task plane, and needs to deviate from the plane in order to reach different locations on or near the plane. Therefore, we mitigate the constraint by allowing the UAVs to move to the spatial points from which the distance to their BS-task planes are within $\Delta$, as the spots shown in Fig. 4.29. The reduced available action set of UAV $i$ at state $\boldsymbol{s}_i = (x_i, y_i, h_i)$ can be defined as follows.

**Definition 4.4** Supposing UAV $i$ at the state $\boldsymbol{s}_i = (x_i, y_i, h_i)$, the location of its task at $\boldsymbol{S}_i = (X_i, Y_i, 0)$, and the location of BS at $\boldsymbol{S}_0 = (0, 0, H_0)$, the action $\boldsymbol{a} = (a_x, a_y, a_h)$ in the reduced available action set $\mathscr{A}_i^+(\boldsymbol{s}_i)$ satisfies the following conditions.

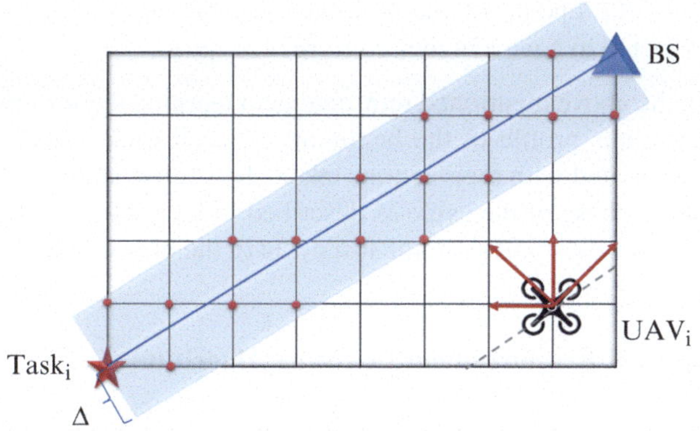


**Fig. 4.29** Illustration on the constrained available action set of UAV $i$

(1) $\text{Dist}(\boldsymbol{s}_i + \boldsymbol{a}; \boldsymbol{S}_i, \boldsymbol{S}_0) \leq \text{Dist}(\boldsymbol{s}_i; \boldsymbol{S}_i, \boldsymbol{S}_0)$ or $\text{Dist}(\boldsymbol{s}_i + \boldsymbol{a}; \boldsymbol{S}_i, \boldsymbol{S}_0) \leq \Delta$;
(2) $x_i + a_x \in [\min(x_i, X_i, 0), \max(x_i, X_i, 0)]$,
$y_i + a_y \in [\min(y_i, Y_i, 0), \max(y_i, Y_i, 0)]$, and
$h_i + a_h \in [h_{\min}, h_{\max}]$.

Here $\text{Dist}(\boldsymbol{s}_i; \boldsymbol{S}_i, \boldsymbol{S}_0)$ denotes the horizontal distance between $s_i$ to the vertical plane passing through $\boldsymbol{S}_i$ and $\boldsymbol{S}_0$.

In Definition 4.4, condition (1) limits the actions to those leading the UAV to the spatial points near the BS-task plane, and condition (2) stops the UAV from moving away from the cross region between the location of its task and the BS.

Moreover, instead of initializing the Q-values for all the possible state-action pairs at the beginning, we propose that the UAVs initialize the Q-values only for the state and the actions within the reduced available action set. In this way, the state sets of the UAVs are reduced to much smaller sets, which makes the learning more efficient.

2. *Model-based Reward Representation*: In both the single-agent Q-learning and the opponent modeling Q-learning algorithms, the UAVs update their Q values based on the information provided by the BS, which indicates the validity of the latest transmitted sensory data. Nevertheless, since the UAVs can only observe the reward to be either 1 or 0, the Q-values converge slowly and the performance of the algorithms is likely to be poor.

Therefore, we propose that the UAVs update their Q-values based on the probabilities of successful valid sensory data transmission. Specifically, UAV $i$ calculates the probability $p_{\text{sTx},i}$ after observing the state-action profile $(\boldsymbol{s}, (a_i, \boldsymbol{a}_{-i}))$ in the latest cycle according to (4.90), and takes it as the reward $R_i$ for the $k$-th cycle.

Moreover, to make the learning algorithm converge with higher speed, in the initialization of the enhanced multi-UAV Q-learning algorithm, we propose that UAV $i$ initializes its $Q_i(\boldsymbol{s}, (a_i, \boldsymbol{a}_{-i}))$ with the calculated $p_{sTx,i}$ for the state-action pair. In this way, the update of the Q-values is more accurate and the learning algorithm is expected to have a higher convergence speed.

*Remark 4.4* In the above mentioned reinforcement learning algorithms, UAVs need to know the locations profile in the beginning of each cycle, and the rewards in the last cycle associated with their actions taken. This information gathering can be done in beaconing phase of the cycle as described in Sect. 4.4.2.1, in which the BS can include the rewards of UAVs in the last cycle in the broadcasted beacon.

#### 4.4.4.4 Analysis of Reinforcement Learning Algorithms

In the final part of this section, we analyze the convergence, the complexity, and the scalability of the proposed reinforcement learning algorithms.

1. *Convergence Analysis*: For the convergence of the reinforcement learning algorithms, it has been proved in [53] that under certain conditions, single-agent Q-learning algorithm is guaranteed to converge to the optimal $Q^*$. In consequence, the policy $\pi$ of the agent converges to the optimal policy $\pi^*$. It can be summarized in the following Theorem 4.16.

**Theorem 4.16** *The Q-learning algorithm given by*

$$\begin{aligned} Q^{(k+1)}(\boldsymbol{s}^{(k)}, \boldsymbol{a}^{(k)}) =& \left(1 - \alpha^{(k)}\right) Q^{(k)}(\boldsymbol{s}^{(k)}, \boldsymbol{a}^{(k)}) \\ &+ \alpha^{(k)} [R(\boldsymbol{s}^{(k)}, \boldsymbol{a}^{(k)}) + \gamma \max_{\boldsymbol{a}'} Q(\boldsymbol{s}^{(k+1)}, \boldsymbol{a}')] \end{aligned} \tag{4.99}$$

*converges to the optimal $Q^*$ values if the following conditions are satisfied.*

1. *The state and action spaces are finite.*
2. $\sum_k \alpha^{(k)} = \infty$ *and* $\sum_k (\alpha^{(k)})^2 < \infty$.
3. *The variance of $R(\boldsymbol{s}, \boldsymbol{a})$ is bounded.*

Therefore, in the multi-agent reinforcement learning cases, if other agents play, or converge to stationary strategies, the single-agent reinforcement learning algorithm also converges to the optimal policy.

However, it is generally hard to prove the convergence with other agents which learn simultaneously. This is because when the agent is learning the Q-value of its actions in the presence of other agents, it faces a non-stationary environment and thus the convergence of Q-values is not guaranteed. The theoretical convergence of the Q-learning in multi-agent cases is guaranteed only in few situations such as in the iterated dominance solvable games and the team games [50]. Like single-agent Q-learning algorithm, the convergence of opponent modeling Q-learning is not generally guaranteed, except for in the setting of iterated dominance solvable games and team matrix game [52].

To handle this problem, we adopt $\alpha^{(k)} = 1/k^{2/3}$ in [54] which satisfies the conditions to achieve the convergence in the single-agent Q-learning, and analyze the convergence of the reinforcement learning in the multi-agent case through simulation results which will be provided in Sect. 4.4.5.

2. *Complexity Analysis*: For the single-agent Q-learning algorithm, the computational complexity in each iteration is $\mathcal{O}(1)$, since the UAV does not consider the other UAVs in the learning process. For the multi-agent Q-learning algorithm, the computational complexity in each iteration is $\mathcal{O}(2^N)$, due to the calculation of the expected discounted reward in (4.97).

   As for the proposed enhanced multi-UAV Q-learning algorithm, each UAV needs to calculate the probability for successful valid data transmission based on Algorithm 11. It can be seen that the recursive Algorithm 11 runs for at most $2^{CT_u}$ times and each iteration has the complexity of $\mathcal{O}(N)$, which makes its overall complexity $\mathcal{O}(N)$. Therefore the complexity of the proposed enhanced algorithm is still $\mathcal{O}(2^N)$, due to the expectation over the joint action space.

   Although the computational complexity of the enhanced multi-UAV Q-learning algorithm in each iteration is in the same order as that of opponent modeling Q-learning algorithm, it reduces the computational complexity and speeds up the convergence by the following means.

   (a) Due to the available action set reduction, the available action set of each UAV is at least reduced to one-half of its original size. This makes the joint action space to be $2^N$ times smaller.
   (b) The reduced available action set leads to a much smaller state space of each UAV. For example, for UAV $i$ and its task at $(X_i, Y_i, 0)$, the original size of its state space can be estimated as $\pi R_{\max}^2(h_{\max} - h_{\min})/\Delta^3$, and the size of its state space after available action set reduction is $2(X_i + Y_i)(h_{\max} - h_{\min})/\Delta^2$, which is $2\Delta/(\pi R_{\max})$ of the original one.
   (c) The proposed algorithm adopts model-based reward representation, which makes the Q-value updating to be more precise, and saves the number of iterations needed to estimate the accurate Q-values of the state-action pairs.

3. *Scalability Analysis*: With the growth of the number of UAVs, the state spaces of the UAVs in the multi-agent Q-learning algorithm and the enhanced multi-UAV Q-learning algorithm grow exponentially. Besides, it can be seen that the enhanced multi-UAV Q-learning algorithm still has exponential computational complexity in each iteration, and thus, it is not suitable for large-scale UAV networks.

To adapt the algorithms to large-scale UAV networks, the reinforcement learning methods need to be combined with function approximation approaches in order to estimate Q-values efficiently. The function approximation approaches take examples from a desired function, Q-function in the case of reinforcement learning, and generalize from them to construct an approximation of the entire function. In this regard, it can be used to estimate the Q-values of the state-action pairs in the entire state space efficiently when the state space is large.

### 4.4.5 Simulation Results

In order to evaluate the performance of the proposed reinforcement learning algorithms for the UAV trajectory design problem, simulation results are presented in this section. Specifically, we use MATLAB to build a frame-level simulation of the UAV sense-and-send protocol, based on the system model described in Sect. 4.4.1 and the parameters in Table 4.6. Besides, the learning ratio in the algorithm is set to be $\alpha^{(k)} = 1/k^{2/3}$ in order to satisfy the converge condition in Theorem 4.16, and the exploration ratio is set to be $\epsilon^{(k)} = 0.8e^{-0.03k}$, which approaches 0 when $k \to \infty$.

Figure 4.30 shows UAV 1's probability of successful valid sensory data transmission versus UAV 1's height and its distance to the BS, given that task 1 is located at $(500, 0, 0)$, and the locations of UAV 2 and UAV 3 are fixed at $(-125, 125, 75), (-125, -125, 75)$, respectively. It can be seen that the optimal point at which UAV 1 has the maximum probability of successful valid sensory data transmission is located in the region between BS and task 1. This is because when the UAV approaches the BS, its successful sensing probability drops, and when the UAV approaches the task, its successful transmission probability suffers. Besides, it is shown that the optimal point for UAV 1 to sense and send is *above*, rather than *on* the BS-task line, where UAV 1 can be closer to both the BS and its task. This is because in the transmission model in Sect. 4.4.1.2, with the increment of the height of the UAV, the LoS probability increases, and thus, the successful uplink transmission probability of the UAV increases.

**Table 4.6** Simulation parameters

| Parameter | Value |
|---|---|
| BS height $H$ | 25 m |
| Number of UAVs $N$ | 3 |
| Noise power $N_0$ | −85 dBm |
| BS decoding threshold $\gamma_{th}$ | 10 dB |
| UAV sensing parameter $\lambda$ | $10^{-3}$/s |
| UAV transmit power $P_u$ | 10 dBm |
| Duration of frame $t_f$ | 0.1 s |
| Distance between adjacent spatial points $\Delta$ | 25 m |
| UAVs' minimum flying height $h_{\min}$ | 50 m |
| UAVs' maximum flying height $h_{\max}$ | 150 m |
| Discount ratio $\rho$ | 0.9 |
| Duration of beaconing phase in frames $T_b$ | 3 |
| Duration of sensing phase in frames $T_s$ | 5 |
| Duration of transmission phase in frames $T_u$ | 5 |

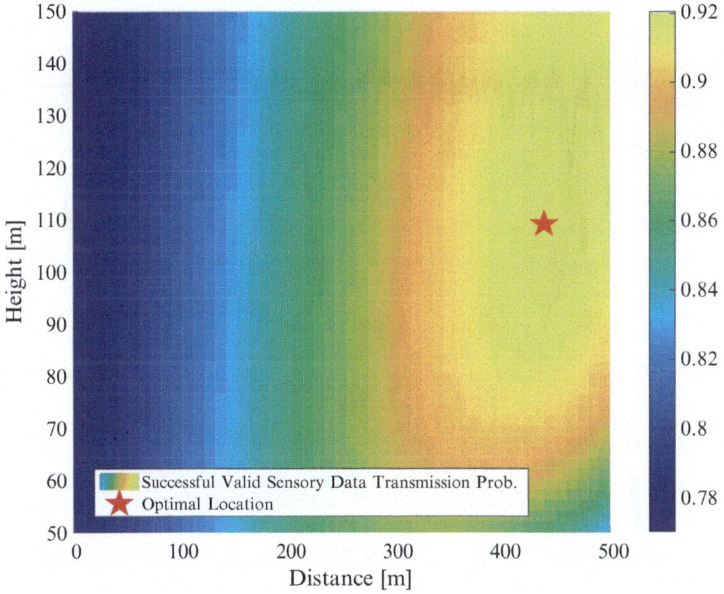


**Fig. 4.30** Successful valid sensory data transmission probability versus the location in the task-BS surface

Figures 4.31 and 4.32 show the average reward per cycle and the average total discounted reward of the UAVs versus the number of cycles in different reinforcement learning algorithm, in which tasks 1–3 are located at (500, 0, 0), $(-250\sqrt{2}, 250\sqrt{2}, 0)$, and $(-250\sqrt{2}, -250\sqrt{2}, 0)$, respectively. It can be seen that compared to the single-agent Q-learning algorithm, the proposed algorithm converges to higher average rewards for the UAVs. This is because the UAV in the enhanced multi-UAV Q-learning algorithm takes the states of all the UAVs into consideration, which makes the estimation of Q-values more accurate. Besides, it can also be seen that compared to the opponent modeling Q-learning algorithm, the proposed algorithm converges faster, due to the available action set reduction and the model-based reward representation.

Moreover, in Fig. 4.33, we can observe that for different distances between the tasks and the BS, the proposed algorithm converges to higher average total discounted rewards for UAVs after 1000 cycles compared to the other algorithms. It can be seen that the average total discounted reward in the three algorithms decreases with the increment of the distance between the BS and the tasks. Nevertheless, the decrement in the proposed algorithm is less than those in the other algorithms. This indicates that the proposed algorithm is more robust to the variance of the tasks' locations than the other two algorithms.

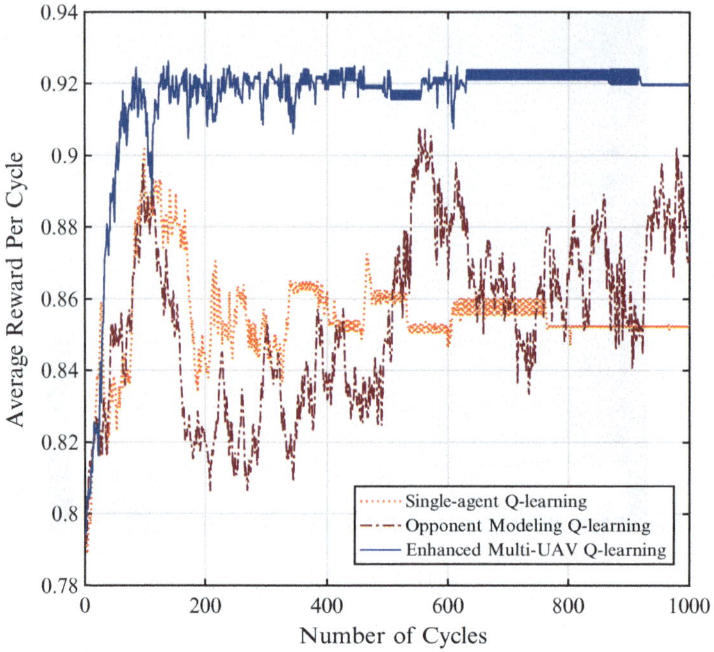


**Fig. 4.31** UAVs' average reward per cycle versus number of cycles of different reinforcement learning algorithms

Figure 4.34 shows the average number of successful valid sensory data transmissions versus the duration of the transmission phase $T_u$ with the proposed algorithm, under different conditions of the distance between the tasks and the BS. It can be seen that the average number of successful valid sensory data transmissions per second first increases and then decreases with the increment of $T_u$. This is because when $T_u$ is small, the successful uplink transmission probability increases rapidly with the increment of $T_u$. However, when $T_u$ is large, the successful uplink transmission probability is already high and increases slightly when $T_u$ becomes larger. Therefore, the average number of successful valid sensory data transmissions per second drops due to the increment of cycles' duration.

Figure 4.35 shows the average number of successful valid sensory data transmissions per second versus $T_u$, under the different conditions of the number of UAVs. It can be seen that when the number of UAVs increases, the average number of successful valid sensory data transmissions per second decreases. This is because the competition among the UAVs for the limited subchannels becomes more intensive. Besides, when the number of UAVs increases, the optimal duration of the transmission phase becomes longer. This indicates that in order to achieve a optimal data rate, the BS needs to increase the duration of the transmission phase when the number of UAVs in the network increases.

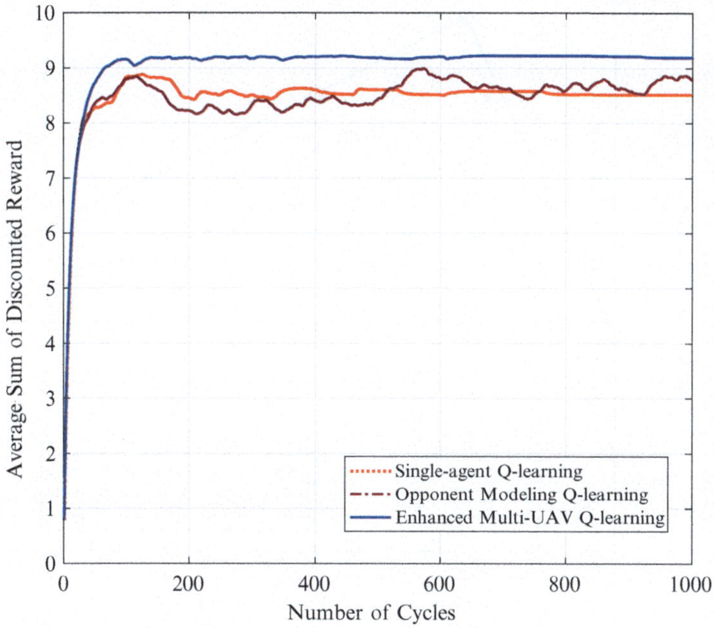


**Fig. 4.32** UAVs' average discounted reward versus number of cycles of different reinforcement learning algorithms

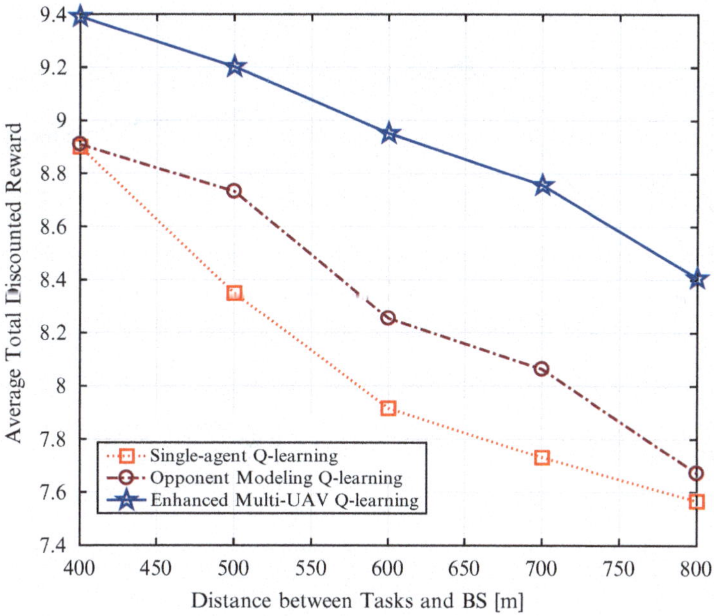


**Fig. 4.33** UAVs' average discounted reward versus distance between tasks and BS in different reinforcement learning algorithms

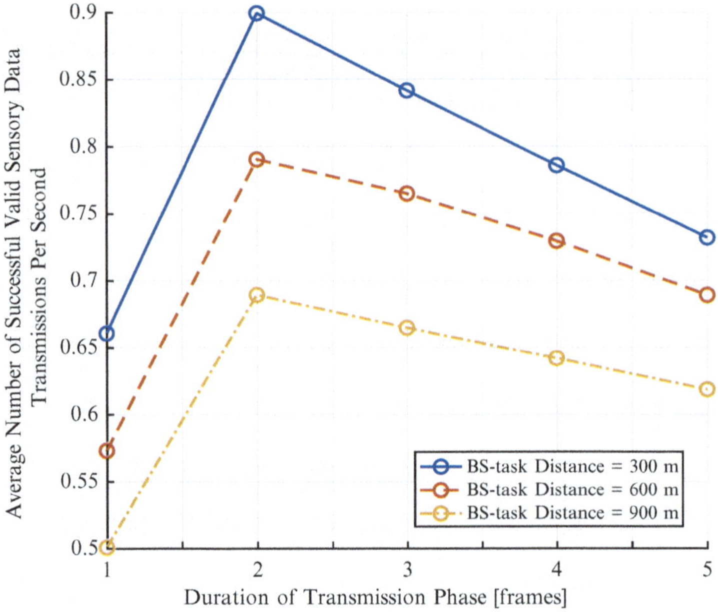


**Fig. 4.34** Average number of successful valid sensory data transmissions per second versus duration of transmission phase $T_u$ under different task distance conditions

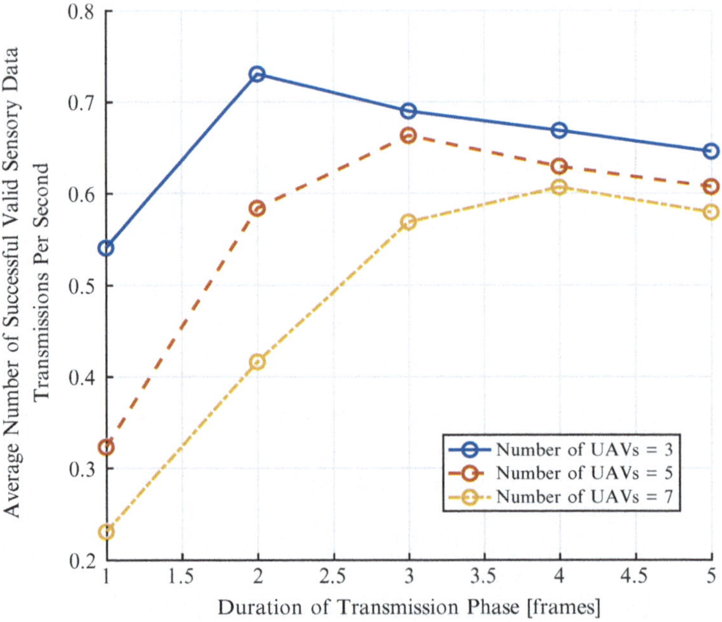


**Fig. 4.35** Average number of successful valid sensory data transmissions per second versus duration of transmission phase $T_u$ under different number of UAVs. Distance between the BS and the tasks = 800 m

### 4.4.6 Summary

In this section, we have considered the scenarios where UAVs performs real-time sensing tasks, and have solved the distributed UAV trajectory design problem by using reinforcement learning. First, we have proposed a decentralized sense-and-send protocol to coordinate multiple UAVs. To evaluate the performance of the protocol, we have solved the probability of successful valid sensory data transmission in the protocol by using nested Markov chains. Then, after the UAV trajectory problem formulation under the reinforcement learning framework, we have proposed the enhanced multi-UAV Q-learning algorithm to solve it efficiently. The simulation results have shown that the proposed algorithm converged faster and achieved higher utilities for the UAVs. We can observe from the simulation results that: 1) our proposed algorithm is more robust to the increment of tasks' distance, compared to single-agent and opponent modeling Q-learning algorithms; 2) moreover, the simulation also have shown that the BS needs to increase the duration of the transmission phase to improve the data rate when the number of UAVs increases.

## 4.5 Applications of the Cellular Internet of UAVs

In a recent report from the World Health Organization [55], air pollution has become the world's largest environmental health risk, as one in eight of global deaths is caused by air pollution exposure each year. Air pollution is caused by gaseous pollutants that are harmful to humans and ecosystem, especially concentrated in the urban areas of developing countries. Thus, reducing air pollution would save millions of lives, and many countries have invested significant efforts on monitoring and reducing the emission of air pollutants. Government agencies have defined AQI to quantify the degree of air pollution. AQI is calculated based on the concentration of a number of air pollutants (e.g., the concentration of $PM_{2.5}$, $PM_{10}$ particles, and so on in developing countries). A higher value of AQI indicates that air quality is "heavily" or "seriously" polluted, resulting in a greater proportion of the population may experience harmful health effects [56]. To intuitively reflect AQI value of locations in either 2D or 3D area, AQI map is defined to offer such convenience [57].

**Mobile AQI Monitoring** AQI monitoring can be completed by sensors at governmental static observation stations, generating an AQI map in a local area (e.g., a city [58]). However, these static sensors only obtain a limited number of measurement samples in the observation area and may often induce high costs. For example, there are only 28 monitoring stations in Beijing. The distance between two nearby stations is typically several ten-thousand meters, and the AQI is monitored every 2 h [59]. To provide more flexible monitoring and reduce the cost, mobile devices, such as cell phones, cars, and balloons are used to carry sensors and process real-time measuring. Crowd-sourced photos contributed by mass of cell phones can help depict the 2D AQI map in a large geographical region in Beijing [60], with a range of 4 km×4 km. Mobile nodes equipped with sensors can provide 100 m ×100 m 2D on-ground concentration maps with relatively high resolution [61–63]. Sensors carried by tethered balloons can build the height profile of AQI at a fixed observation height within 1000 m [64]. A mobile system with sensors equipped in cars and drones can help monitor $PM_{2.5}$ in open 3D space [65], with 200 m per measurement.

**Motivations for Real-Time Fine-Grained Monitoring** Even though current mobile sensing approaches can provide relatively accurate and real-time AQI monitoring data, they are spatially coarse-grained, since two measurements are separated by few hundreds of meters in horizontal or vertical directions in the 3D space. However, AQI has intrinsic changes from meters to meters, and it is preferred to perform AQI monitoring in the 3D space surrounding an office building or throughout a university campus, rather than city-wide [66, 67]. The AQI distribution in meter-sliced areas, called as fine-grained areas would be desirable for people, particularly those living in urban areas. The fine-grained AQI map can help design the ventilation system for buildings, which, for example, can guide teachers and students to stay away from the pollution sources on campus [68].

Due to the high power consumption of mobile devices, one can only measure a limited number of locations of the entire space. To avoid an exhaustive measurement, using an estimation model to approximate the value of unmeasured area has been widely adopted. In [69], the prediction model is based on a few public air quality stations and meteorological data, taxi trajectories, road networks, and Point of Interests (POIs). However, because they estimate AQI using a feature set based on historical data, their model cannot respond in real time to the change in pollution concentration at an hourly granularity, leading to large errors at times. In [65], the random walk model is used for prediction by dividing the whole space into different shapes of cubes. However, the model may not reflect physical dispersion of particles [70, 71], and all locations are measured without considering the battery life constraint when mobile devices are used. Mobile sensor nodes used in [61] employ the regression model as well as graph theory to estimate the AQI value at unmeasured locations. However, they mainly focus on 2D area, and can hardly produce a 3D fine-grained map. Neural networks (NN) are also used for forecasting on the AQI distribution [72–75]. However, its performance in fine-gained area is not satisfied without considering the physical characteristic of real AQI distribution.

In this section, we design a mobile sensing system based on UAVs, called *ARMS*, that can effectively catch AQI variance at meter-level and profile the corresponding fine-grained distribution, as illustrated in Fig. 4.36. ARMS is a real-time monitoring system that can generate current AQI map within a few minutes, compared to the previous methods with an interval of a few hours. With ARMS, the fine-grained AQI map construction can be decomposed into two parts. First, we propose a novel AQI distribution model, named Gaussian Plume model embedding Neural Networks (GPM-NN), that combines physical dispersion and non-linear NN structure, to do predictions of unmeasured area. Second, we detail the adaptive

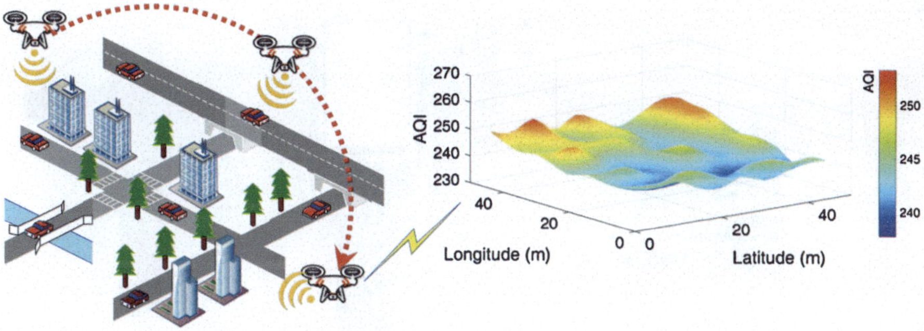


**Fig. 4.36** An illustration of AQI measurement using mobile sensing over UAV

monitoring algorithm as well as address its applications in a few typical scenarios. By measuring only selected locations in different scenarios, GPM-NN is used to estimate AQI value at unmeasured locations and generate real-time AQI maps, which can save the battery life of mobile devices while maintaining high accuracy in AQI estimation.

The rest of this section is organized as follows. In Sect. 4.5.1, we briefly introduce our UAV sensing system. In Sect. 4.5.2, we present our fine-grained AQI distribution model. The adaptive monitoring algorithm is addressed in Sect. 4.5.3. In Sects. 4.5.4 and 4.5.5, we present two typical application scenarios and performance analysis of ARMS, respectively. Finally, this section is summarized in Sect. 4.5.6.

## *4.5.1 Preliminaries of UAV Sensing System*

In this subsection, we first provide a brief introduction of ARMS [5], and then we show how to construct a dataset using ARMS. To confirm the reliability of the collected dataset, we compare the collected data and the official AQI measured by the nearest Beijing government's monitoring station, i.e., the Haidian station [76]. To determine the parameters of our model, we test possible factors that may influence AQI, such as wind, locations, etc., and remove those factors that have small correlations with AQI in the fine-grained scenarios from our model.

### 4.5.1.1 System Overview

The architecture of ARMS includes an UAV and an air quality sensor boarded on the UAV, as shown in Fig. 4.37. The sensor is fixed in a plastic box with vent holes, bundled on the bottom of UAV. The sensor uses a laser-based AQI detector [77], which can provide the concentration within $\leq \pm 3\%$ monitor error for common pollutants in AQI calculation, such as $PM_{2.5}$, $PM_{10}$, CO, NO, $SO_2$, and $O_3$. The

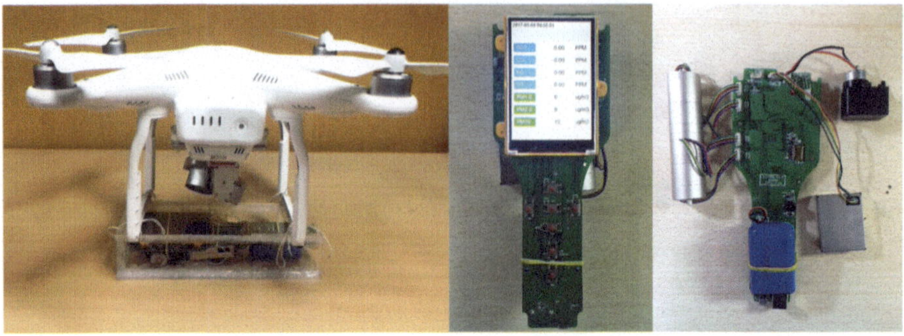

**Fig. 4.37** The ARMS system, the front and the back of the sensor board

values of these pollutants are real-time recorded, with which we calculate the corresponding AQI value at measuring locations.

For the UAV, we select DJI Phantom 3 Quadcopter [78] as the mobile sensing device. The UAV can keep hosting for at most 15 min due to the battery constraint, which restricts the longest continuous duration within one measurement. The GPS sensor on the UAV can provide the real-time 3D position. During one measurement, the UAV is programmed with a trajectory, including all locations that need to be measured. Following this trajectory, UAV hovers for 10 s to collect sufficient data to derive the AQI value at each stop, before moving to the next one.

During one monitoring process, ARMS measures all target locations and records the corresponding AQI values. After the measuring process, the data is then sent to the offline PC and put into the GPM-NN model to construct the real-time AQI map. Thus, the map construction process is offline.

#### 4.5.1.2 Dataset Description

Data collected by ARMS are then arranged as a dataset.[9] As shown in Fig. 4.36, we have conducted a measurement study in both typical 2D and 3D scenarios (i.e., a roadside park and the courtyard of an office building in Peking University), respectively, from Feb. 11 to Jul. 1, 2017, for more than 100 days to collect sufficient data [79].

In the dataset, each *.txt* file includes one complete measurement over a day in one typical scenario. In each *.txt* file, each sample has four parameters, 3D coordinates $(x, y, z)$ and an AQI value. Each value represents the measured AQI, while its coordinates in the matrix reflect the position in different scenarios. In the 2D scenario, we assume $z = 0$, while measuring at an interval of 5 m in $x$ and $y$

[9]Dataset can be found at https://github.com/YyzHarry/AQI_Dataset.

directions. In the 3D scenario, every row presents fixed position in $xy$ plane, while every column represents the height at an interval of 5m in $z$ direction.

#### 4.5.1.3 Data Reliability

To verify that there is no measurement error, we show the results of the relationship between our collected data and the official data (i.e., Haidian station [76]), in Fig. 4.38. Note that the official data is limited and only for the 2D space, while our system is mobile and suitable for the 3D space profiling. We select 14 consecutive days for about 60 instances of monitoring from Mar. 14 to Mar. 27, 2017, to verify the reliability of our measurement. We use the two-tailed hypothesis test [80]: $\mathscr{H}_0 : \mu_1 = \mu_2 \ vs. \ \mathscr{H}_1 : \mu_1 \neq \mu_2$, where $\mu_1$ denotes our average measured data for all days and $\mu_2$ is the average for the official ones. The test result, $P = 0.9999 \gg 0.05$, indicates that there is no significant difference between the two values, which confirms the reliability of our measurements.

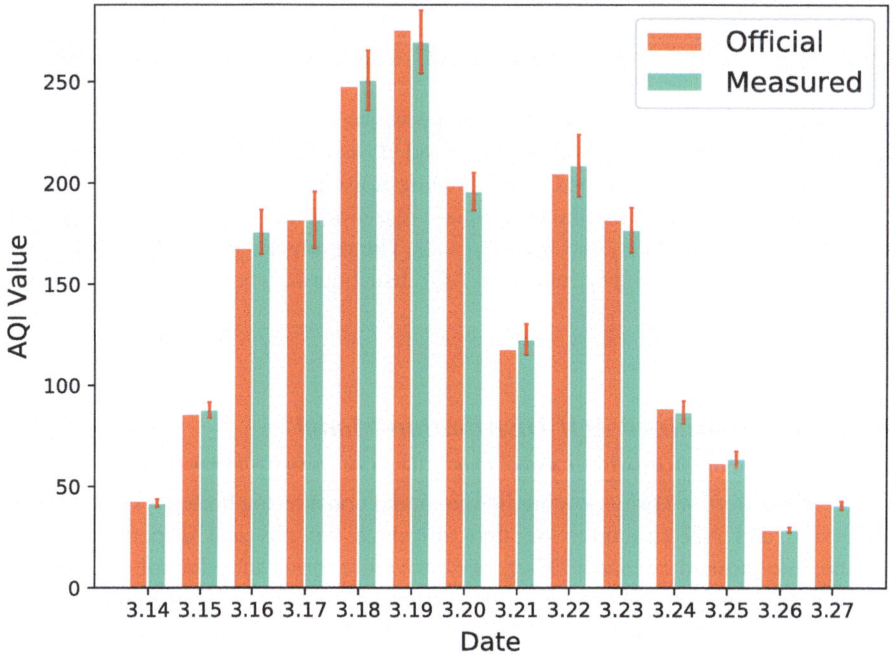


**Fig. 4.38** AQI value comparison between official data and data we collected, for 14 days in March, 2017

**Table 4.7** Result of the hypothesis test

| Tested parameter | P_value |
|---|---|
| Wind | $7.5693 \times 10^{-5}$ ($\ll 0.05$) |
| Location | $2.0981 \times 10^{-5}$ ($\ll 0.05$) |
| Temperature | 0.9070 ($\gg 0.05$) |
| Humidity | 0.6996 ($\gg 0.05$) |

#### 4.5.1.4 Selection of Model Parameters

According to the previous AQI monitoring results for coarse-grained scenarios [71], AQI is related to wind (including speed and direction), temperature, humidity, altitude and spatial locations. But for fine-grained scenarios, correlations between AQI and these spatial parameters need to be reconsidered, due to the heterogeneous diffusion in both vertical and horizontal directions in a small-scale area. In this test, all these potential parameters are measured by our ARMS with different sensors. To evaluate the correlation between these parameters, we adopt the spatial regression according to [81], and test the coefficient for each parameter. Mathematically, the spatio-temporal model is given below:

$$C(s_i) = z(s_i)\boldsymbol{\beta}^{\mathrm{T}} + \varepsilon(s_i), \tag{4.100}$$

where $C(s_i)$ is the particle concentration at position $s_i$, $z(s_i) = (z_1(s_i), \ldots, z_n(s_i))$ denotes the vector of $n$ parameters at $s_i$, and $\boldsymbol{\beta} = (\beta_1, \ldots, \beta_n)$ is the coefficient vector. $\varepsilon(s_i) \sim N(0, \sigma^2)$ is the Gaussian white-noise process.

Based on our data, we use the least square regression and implement a hypothesis test for each coefficient $\beta_j$, as $\mathscr{H}_0 : \beta_j = 0$. The results in Table 4.7 indicate that wind and location are highly related to AQI distribution, whereas temperature and humidity are not.

### 4.5.2 *Fine-Grained AQI Distribution Model*

In this subsection, we provide a prediction model considering both physical particle dispersion and NN structure. We first introduce the physical dispersion model for the fine-grained scenario. Then, we provide a brief introduction of NN we adopt in modeling, which can adapt to complicated cases, such as the non-linearity introduced by extreme weather. Finally, we embed the dispersion model in NN to design our model.

#### 4.5.2.1 Physical Particle Dispersion Model

We first address the physical particle dispersion model for fine-grained scenarios. Specifically, we ignore the influence of temperature and humidity according to discussions in Section 2.D, and select the Gaussian Plume Model (GPM) in the particle movement theory [82], to describe the particle's dispersion. GPM is widely used to describe particles' physical motion [70, 83], and its robustness has been proved in a small scale system [84]. GPM is expressed as

$$C(x, y, z) = \frac{Q}{2\pi \sigma_y \sigma_z u} \exp\left(-\frac{(z-H)^2}{2\sigma_z^2}\right) \exp\left(-\frac{y^2}{2\sigma_y^2}\right), \tag{4.101}$$

where $Q$ is the point source strength, $u$ is the average wind speed, and $H$ denotes the height of source.

To adopt GPM into the fine-grained scenario, the GPM is revised as below

$$\begin{aligned} C(\boldsymbol{x}, u) &= \int_{-\frac{L}{2}}^{\frac{L}{2}} \frac{\lambda}{2\pi \sigma_y \sigma_z u} \exp\left(-\frac{(z-H)^2}{2\sigma_z^2} - \frac{y^2}{2\sigma_y^2}\right) \mathrm{d}y \\ &= \frac{\lambda \exp\left(-\frac{(z-H)^2}{2\sigma_z^2}\right)}{2\pi \sigma_z u} \int_{-\frac{L}{2\sigma_y}}^{\frac{L}{2\sigma_y}} \exp\left(-\frac{\gamma^2}{2}\right) \mathrm{d}\gamma \\ &= \frac{\lambda}{\sqrt{2\pi}\sigma_z u} \exp\left(-\frac{(z-H)^2}{2\sigma_z^2}\right) \left[1 - 2Q\left(\frac{L}{2\sigma_y}\right)\right], \end{aligned}$$

where $C(\boldsymbol{x}, u)$ is the AQI value at location $\boldsymbol{x}$, $u$ is the real wind speed at different locations in the entire space, and $H$ denotes a variable that reflects the influence of wind direction, which presents severely polluted areas along $z$-axis. Pollution mainly derives as a line source aligned the $y$-axis, and $L$ denotes the length of polluted source, $\lambda$ denotes the particle density at the source. $\sigma_y$ and $\sigma_z$ are diffusion parameters in $y$ and $z$ directions, and are both empirically given. The dispersion model in (4.102) can reflect physical characteristics, but can hardly deal with unpredictable complicated changes, such as the non-linearity introduced by extreme weather.

#### 4.5.2.2 Neural Network Model

The neural network model, especially multilayer perceptron (MLP), has been widely adopted to do estimation for air quality [72–75]. They usually train models by using a huge amount of data to achieve decent performance. All possible influential factors are involved as the neural network input variables for network training. Other types of NN [85, 86] are proposed for better classification with more complex structures. As it has been proved that a three-layer neural network can compute any arbitrary function [87–89], NN is able to present the complicated changes in fine-grained scenario. However, without considering the physical characteristics of AQI, the NN model may overfit and perform worse on the test data than on the training data [72].

### 4.5.2.3 GPM-NN Model

In order to utilize the advantages of both GPM and NN, we embed the revised GPM in NN, and put forward GPM-NN model.

1. *Model Description*: As shown in Fig. 4.39, the model structure contains a linear part (the physical dispersion model) and a non-linear part (the NN structure) for fine-grained AQI distribution, respectively. Let $N$ be the total number of data collected by ARMS, which is represented by a pair $(\boldsymbol{X}_j, t_j)$, where $\boldsymbol{X}_j = [x_1\ x_2\ \ldots\ x_m]^{\mathrm{T}}$ is the $j$th sample with a dimensionality of $m$ variables and $t_j$ is the measured AQI value.
   (a) In the non-linear NN part, let $K$ denote the total number of neurons in the hidden layer. The weights for these neurons are denoted by $\boldsymbol{W} = [\boldsymbol{W}_1\ \boldsymbol{W}_2\ \ldots\ \boldsymbol{W}_K]$, where $\boldsymbol{W}_i = [\omega_{i1}\ \omega_{i2}\ \ldots\ \omega_{im}]$ is the $m$-dimensional weight vector containing the weights between the components of input vectors and the $i$th neuron in the hidden layer. $\boldsymbol{b} = [b_1\ b_2\ \ldots\ b_K]$ is the bias term of the $i$th neuron. The non-linear part with $K$ neurons in the hidden layer will have $\boldsymbol{\beta} = [\beta_1\ \beta_2\ \ldots\ \beta_K]$ as weights for output layer and $g(\cdot)$ is the activation function.
   (b) In the linear part, we use $C(\boldsymbol{x}, u)$, a constant value and a Gaussian process as inputs, to reflect the influence of the physical model. The regression weights are correspondingly determined as $\beta_{K+1}$, $\beta_{K+2}$, and 1.

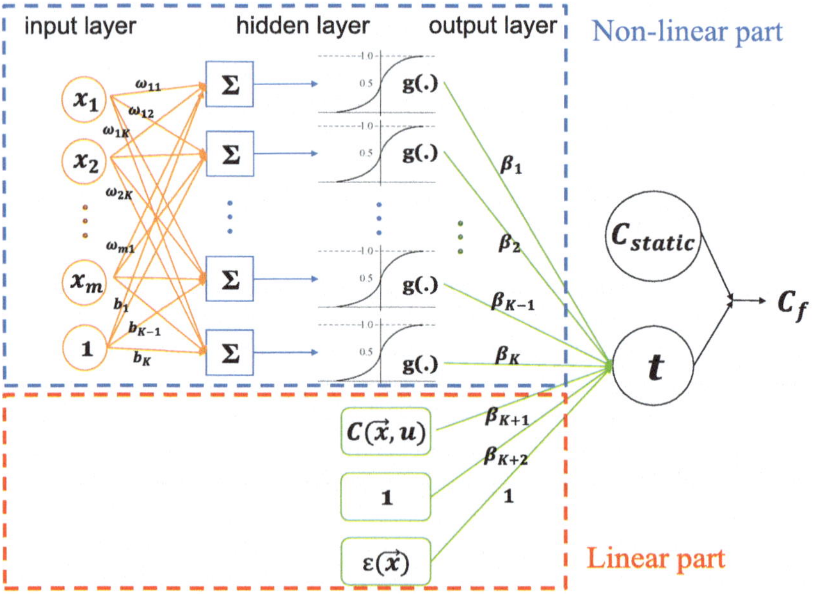


**Fig. 4.39** The model structure of GPM-NN

Thus, the mathematical expression of the proposed model can be written as

$$
\begin{aligned}
t(\boldsymbol{x}, u) = & \sum_{i=1}^{K} \beta_i g(\boldsymbol{W}_i \boldsymbol{X}_j + b_i) + \beta_{K+1} C(\boldsymbol{x}, u) + \\
& \beta_{K+2} + \varepsilon(\boldsymbol{x}), \qquad j = 1, 2, \ldots, N,
\end{aligned}
\tag{4.102}
$$

where $t(\boldsymbol{x}, u)$ is the estimated value of $t_j$ and it represents the model's output. $C(\boldsymbol{x}, u)$ is the output of the dispersion model in (4.102) and $\beta_i$ are regression coefficients. $\varepsilon(\boldsymbol{x}) \sim N(0, \sigma^2)$ is the measurement error defined by a Gaussian white-noise process. Since there is a risk that the NN part will overfit and perform worse on the test data than training data, the estimated AQI value is expressed as

$$
C_f(\boldsymbol{x}, u) = C_{static} + t(\boldsymbol{x}, u), \tag{4.103}
$$

where $C_{static}$ is the average value of our measured AQI in a day, which is an invariant to quantify basic distribution characteristics.

2. *Parameter Estimation*: As shown in (4.102), GPM-NN has $(K{+}3)$ parameters, $H, \beta_1, \beta_2, \ldots, \beta_{K+2}$, which need to be estimated based on the data collected by ARMS. 50 days' data are used for training the non-linear part of GPM-NN. We use the least square regression to estimate the parameters. Let $\mathscr{S}$ denote the residual error as

$$
\mathscr{S} = \sum_{i=1}^{N} \left\| \hat{C}_f(\boldsymbol{x}_i, u_i) - \beta_{K+2} - \beta_{K+1} C(\boldsymbol{x}, u) - \sum_{j=1}^{K} \beta_j g_j \right\|^2, \tag{4.104}
$$

where $i$ denotes the measuring sample of the $i$th observation point, and $g_j = g(\boldsymbol{W}_j \boldsymbol{X}_i + b_j)$.

**Proposition 4.2** *Equation (4.104) has a unique minimum point for estimated parameters* $\beta_1, \beta_2, \ldots, \beta_{K+2}$ *and* $H$, *when* $\sigma_z^2 > \max\{2z_i^2, 2H_0^2\}$.

***Proof*** For $\beta_j$ where $j \in [1, K{+}2]$, we have

$$
\frac{\partial^2 \mathscr{S}}{\partial \beta_j^2} = \begin{cases} 2\sum_{i=1}^{N} g_j^2 > 0, & 1 \le j \le K, \\ 2\sum_{i=1}^{N} C^2(\boldsymbol{x}_i, u_i) > 0, & j = K+1, \\ 2\sum_{i=1}^{N} 1 = 2N > 0, & j = K+2. \end{cases} \tag{4.105}
$$

Hence, $\partial \mathscr{S} / \partial \beta_j$ are all convex functions, with $j \in [1, K{+}2]$.

As for variable $H$, the second order partial derivative can be calculated as

$$\frac{\partial^2 \mathscr{S}}{\partial H^2} = -2\sum_{i=1}^{N} \beta' \Bigg[ -\frac{\beta'}{\sigma_z^6 u_i^2}(z_i - H)^2 \exp\left(-\frac{(z_i-H)^2}{\sigma_z^2}\right) - \left(C - \frac{\beta' \exp\left(-\frac{(z_i-H)^2}{2\sigma_z^2}\right)}{u_i \sigma_z}\right) \frac{1}{u_i \sigma_z^3} \exp\left(-\frac{(z_i-H)^2}{2\sigma_z^2}\right) + \left(C - \frac{\beta' \exp\left(-\frac{(z_i-H)^2}{2\sigma_z^2}\right)}{u_i \sigma_z}\right) \frac{(z_i-H)^2 \exp\left(-\frac{(z_i-H)^2}{2\sigma_z^2}\right)}{u_i \sigma_z^5} \Bigg], \tag{4.106}$$

where $C = \left(\hat{C}_f(\boldsymbol{x}_i, u_i) - C_{static} - \beta_{K+2} - \sum_{j=1}^{K} \beta_j g_j\right)$, and $\beta' = \frac{\lambda}{\sqrt{2\pi}} \beta_{K+1} \left(1 - 2Q\left(\frac{L}{2\sigma_y}\right)\right)$. Then we have

$$\frac{\partial^2 \mathscr{S}}{\partial H^2} = 2\sum_{i=1}^{N} \Bigg[ \left(\frac{2\beta'^2 (z_i - H)^2}{\sigma_z^6 u_i^2} - \frac{\beta'^2}{\sigma_z^4 u_i^2}\right) \exp\left(-\frac{(z_i - H)^2}{\sigma_z^2}\right) + C\left(\frac{\beta'}{\sigma_z^3 u_i} - \frac{\beta'(z_i - H)^2}{u_i \sigma_z^5}\right) \exp\left(-\frac{(z_i - H)^2}{2\sigma_z^2}\right)\Bigg]. \tag{4.107}$$

Let $t_i = \exp\left(-\frac{(z_i-H)^2}{2\sigma_z^2}\right)$, each item of the summation is equivalent to a quadratic function $Q_i(t_i) = a_i t_i^2 + b_i t_i$. Note that $t_i \in (0, 1]$, and $t_i = 0$ is one zero point of $Q_i(t_i)$. To satisfy the proposition that $\partial^2 \mathscr{S}/\partial H^2$ always has positive value, the problem becomes

$$\begin{cases} a_i = \dfrac{2\beta'^2 (z_i - H)^2}{u_i^2 \sigma_z^6} - \dfrac{\beta'^2}{u_i^2 \sigma_z^4} < 0, \\ b_i = C\left(\dfrac{\beta'}{u_i \sigma_z^3} - \dfrac{\beta'(z_i - H)^2}{u_i \sigma_z^5}\right) > 0, \end{cases} \quad \forall i \in [1, N], \tag{4.108}$$

which can be simplified as:

$$\begin{cases} \sigma_z^2 > \max\limits_i \, 2(z_i - H)^2, \\ \sigma_z^2 > \max\limits_i \, (z_i - H)^2. \end{cases} \tag{4.109}$$

We define $H \in [0, H_0]$, where $H_0$ is the upper bound for a fine-grained measurement. Hence, by choosing appropriate diffusion parameter $\sigma_z$ as $\sigma_z^2 > \max\{2z_i^2, 2H_0^2\}$, we have

$$\frac{\partial^2 \mathcal{S}}{\partial H^2} = 2\sum_{i=1}^{N} Q_i(t_i) = 2\sum_{i=1}^{N}(a_i t_i^2 + b_i t_i)$$
$$= 2\sum_{i=1}^{N}\left(\frac{b_i^2}{4|a_i|} - |a_i|\left(t_i + \frac{b_i}{2a_i}\right)^2\right) > 0,\ \forall t_i \in (0, 1]. \tag{4.110}$$

Therefore, $\partial\mathcal{S}/\partial H$ is also a convex function, which indicates that Eq. (4.104) has a minimum as well as a unique value, correspondingly. □

To find the minimum point of the residual error function $\mathcal{S}(H, \beta_1, \ldots, \beta_{K+2})$, we use the Newton method [90] to solve the following equations whose analytical solution does not exist, as

$$\begin{cases} \dfrac{\partial \mathcal{S}}{\partial H} = 0, \\ \dfrac{\partial \mathcal{S}}{\partial \beta_j} = 0, \qquad j = 1, 2, \ldots, K+2. \end{cases} \tag{4.111}$$

When the estimation value of $H$ (denoted as $H^*$) is determined, $C(\boldsymbol{x}, u)$ is correspondingly determined. Denote

$$\boldsymbol{J} = \begin{bmatrix} g(\mathbf{W}_1\boldsymbol{X}_1+b_1) & \cdots & g(\mathbf{W}_K\boldsymbol{X}_1+b_K) & C(\boldsymbol{x}_1,u_1) & 1 \\ g(\mathbf{W}_1\boldsymbol{X}_2+b_1) & \cdots & g(\mathbf{W}_K\boldsymbol{X}_2+b_K) & C(\boldsymbol{x}_2,u_2) & 1 \\ \vdots & \ddots & \vdots & \vdots & \vdots \\ g(\mathbf{W}_1\boldsymbol{X}_N+b_1) & \cdots & g(\mathbf{W}_K\boldsymbol{X}_N+b_K) & C(\boldsymbol{x}_N,u_N) & 1 \end{bmatrix}_{N\times(K+2)}$$

as the model output matrix, and similarly

$$\boldsymbol{\beta} = \begin{bmatrix} \beta_1 \\ \beta_2 \\ \vdots \\ \beta_K \\ \beta_{K+1} \\ \beta_{K+2} \end{bmatrix}_{(K+2)\times 1}$$

is the vector that needs to be estimated. Hence, the estimated value of $N$ samples can be written as

$$\boldsymbol{T} = \boldsymbol{J}\boldsymbol{\beta}. \tag{4.112}$$

Note that $\boldsymbol{J}$ is both row-column full rank matrix, which has a corresponding generalized inverse matrix [91]. As we have proved, (4.104) has a unique minimum point, we then have

$$\begin{aligned}\boldsymbol{\beta} &= (\boldsymbol{J}^{\mathrm{T}}\boldsymbol{J})^{-1}\boldsymbol{J}^{\mathrm{T}}\boldsymbol{J}\boldsymbol{\beta} \\ &= (\boldsymbol{J}^{\mathrm{T}}\boldsymbol{J})^{-1}\boldsymbol{J}^{\mathrm{T}}\boldsymbol{T} \\ &= \boldsymbol{J}^{\dagger}\boldsymbol{T},\end{aligned} \tag{4.113}$$

where $\boldsymbol{J}^{\dagger} = (\boldsymbol{J}^{\mathrm{T}}\boldsymbol{J})^{-1}\boldsymbol{J}^{\mathrm{T}}$ is known as the Moore–Penrose pseudo inverse of $\boldsymbol{J}$. This equation is the least squares solution for an over-determined linear system and is proved to have the unique minimum solution [92]. Thus, this equation is equal to the multivariate equation in (4.111), by which we can find the minimum value point of $\mathscr{S}$.

3. *Performance Evaluation*: To determine the initial value of the weights $\boldsymbol{W}$ and biases $\boldsymbol{b}$ for the hidden layer, we use the training data to do preprocessing and acquire the optimal values. Hence, the model can be completely determined for describing the AQI distribution in fine-grained scenarios.

For evaluating the performance of GPM-NN, we use average estimation accuracy (AEA) as the merit, expressed as

$$\overline{AEA} = \frac{1}{n}\sum_{i=1}^{n}\left(1 - \frac{|\hat{C}_f(i) - C_f(i)|}{C_f(i)}\right), \tag{4.114}$$

where $n$ denotes the total locations in the scenario, $\hat{C}_f(i)$ denotes the estimation AQI value in the $i$th location, and $C_f(i)$ denotes the real measured value. In Sects. 4.5.4 and 4.5.5, we compare the accuracy of AQI map constructed by our GPM-NN and other existing models.

### 4.5.3 Adaptive AQI Monitoring Algorithm

In this subsection, we provide the adaptive monitoring algorithm of ARMS. Intuitively, a larger number of measurement locations introduce a higher accuracy of the AQI map. However, based on the physical characteristic of particle dispersion in GPM-NN, we can build a sufficiently accurate AQI map by regularly measuring only a few locations. This process can effectively save the energy, and thus improve the efficiency of the system. Specifically, an AQI monitoring is decomposed into two steps—*complete monitoring* and *selective monitoring*—for efficiency and accuracy. An example of these two steps is shown in Fig. 4.40. We first trigger *complete monitoring* everyday for one time, to establish a baseline distribution. Then ARMS periodically (e.g., every 1 h) measures only a small set of observation points, which

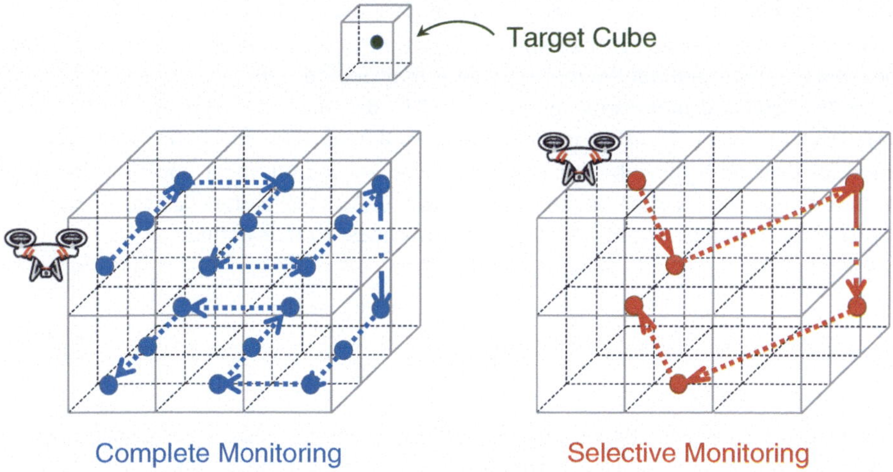


**Fig. 4.40** An example of the adaptive monitoring algorithm, i.e., complete and selective monitoring

are acquired by analyzing the characteristic of the established AQI map. This process, named as *selective monitoring*, is based on GPM-NN to update the real-time AQI map. By accumulating current measurements with the previous map, a new AQI map is generated timely. Every time when selective monitoring is done, ARMS compares the newly measured results and the most recent measurement. If there is a large discrepancy between them, which indicates that the AQI experiences severe environmental changes, we would again trigger the complete monitoring to rebuild the baseline distribution. Thus, ARMS can effectively reduce the measurement effort as well as cope with the unpredictable spatio-temporal variations in the AQI values.

#### 4.5.3.1 Complete Monitoring

The *complete monitoring* is designed to obtain a baseline characteristic of the AQI distribution in a fine-grained area and is triggered at a day interval.

The entire space can be divided into a set of $5\,\mathrm{m} \times 5\,\mathrm{m} \times 5\,\mathrm{m}$ cubes. In the complete monitoring process, ARMS measures all cubes continuously and builds a baseline AQI map using GPM-NN. The process is of high dissipation, and thus is triggered over a long observation period.

#### 4.5.3.2 Selective Monitoring

To reflect changes of the AQI distribution in a small-scale space over time (e.g., between each hour in a day) [65], ARMS uses the *selective monitoring* to capture such dynamics. The selective monitoring makes use of the previous AQI map, by analyzing the physical characteristics of it, to reduce the monitoring overhead in the next survey and maintain the real-time AQI map accordingly.

In the selective monitoring process, ARMS measures the AQI value of only a small set of selected cubes and generates AQI map over the entire fine-grained area. To deal with the inherent trade-off between measurement consumption and accuracy, we put forward an important index called the partial derivative threshold (*PDT*), to guide system selecting specific cubes. *PDT* is defined as

$$PDT_i = \frac{\left|\frac{\partial C_f}{\partial x_i}\right| - \left|\frac{\partial C_f}{\partial x_i}\right|_{min}}{\left|\frac{\partial C_f}{\partial x_i}\right|_{max} - \left|\frac{\partial C_f}{\partial x_i}\right|_{min}}, \tag{4.115}$$

where $x_i$ denotes the $i$th variable in GPM-NN ($i = 1, 2, \ldots, m$), and $C_f = C_f(\boldsymbol{x}, u)$ denotes the entire distribution in a small-scale area. $|\partial C_f/\partial x_i|_{min}$ and $|\partial C_f/\partial x_i|_{max}$ denote the minimum and the maximum value of the partial derivative for parameter $x_i$, respectively. Note that $\partial C_f/\partial x_i$ describes the upper bound of dynamic change degrees we can tolerate, expressed as

$$\frac{\partial C_f}{\partial x_i} = PDT_i \cdot \left(\left|\frac{\partial C_f}{\partial x_i}\right|_{max} - \left|\frac{\partial C_f}{\partial x_i}\right|_{min}\right) + \left|\frac{\partial C_f}{\partial x_i}\right|_{min},$$
$$0 \le PDT_i \le 1. \tag{4.116}$$

For each parameter, there is one corresponding *PDT*. In general, *PDT* reflects the threshold for dynamic change degrees in a fine-grained area. Area that has large change rate of model's parameters would have a larger *PDT* value, indicating more drastic changes. When given a specific *PDT*, any cube whose $\partial C_f/\partial x_i$ is above threshold of (4.116) will be moved into a set $\mathscr{M}$. Moreover, when $PDT_i$ is too small (less than a small const $\delta$), the corresponding $i$th cube will also be added into $\mathscr{M}$. Mathematically, set $\mathscr{M}$ is given as

$$\mathscr{M} = \{i \mid PDT_i \ge PDT\} \cup \{i \mid PDT_i \le \delta\}. \tag{4.117}$$

*Remark 4.5* Elements in $\mathscr{M}$ can be the severe changing areas in a small-scale space (e.g., a tuyere or abnormal building architecture), or typically the lowest or the highest value that can reflect basic features of the distribution. These elements are sufficient to depict the entire AQI map, and hence are needed to be measured between two measurements. Thus, by only measuring cubes in $\mathscr{M}$, ARMS can generate a real-time AQI map implemented by GPM-NN, while greatly reducing the measurement overhead.

In general, *PDT* is adjusted manually for different scenarios. When *PDT* is low, the threshold for abnormal cubes declines, indicating the measuring cubes will increase and the estimation accuracy is relatively high. However, it can cause a great battery consumption. On the other hand, as *PDT* is high, the measuring cubes will decrease. This can cause a decline in accuracy, but can highly reduce consumption. In summary, the trade-off between accuracy and consumption should be studied to acquire a better performance of whole system.

#### 4.5.3.3 Trajectory Optimization

When target cubes in set $\mathscr{M}$ are determined, the total network can be modeled as a 3D graph $G = (V, E)$ with a number of $|V|$ target cubes. Hence, finding the minimum trajectory over these cubes is equal to find the shortest Hamiltonian cycle in a 3D graph. This problem is known as the traveling salesman problem (TSP), which is NP-hard [93].

To solve TSP in this case, we propose a greedy algorithm to find the sub-optimal trajectory. In the fine-grained scenario, ARMS can monitor no more than $n$ cubes over one measurement due to the limited battery. To find the corresponding trajectory, we focus on how to determine the next measuring cube based on current location of ARMS. Let $\mathbb{Z} = \{O_0, O_1, \ldots, O_{|V|-1}\}$ be the set of coverage cubes, with $O_i$ being every observation cube. The aim is to acquire as many target cubes as possible over the trajectory for a higher AQI estimation accuracy. Considering the significant physical characteristic of *PDT* above, we aim to maximize the next cube's *PDT*, as well as minimize the traveling cost from current location to next cube. Hence, finding the optimal trajectory in this case is equal to an iteration of solving the following optimization problem, expressed as

$$\begin{aligned} i^* &= \arg\max_i \left| \frac{PDT_i}{cost(i)} \right| \\ s.t. \quad & O_i \in \mathscr{M}, \\ & O_i \cap \bigcup \{O_0, O_1, \ldots, O_{i-1}\} = \varnothing, \end{aligned} \tag{4.118}$$

where $cost(i)$ is the consumption for the UAV to traverse from the $(i-1)$th cube to the $i$th cube, and $PDT_i$ is acquired by analyzing the characteristic of latest AQI map.

For every current location $i$, the selection of next target cube follows (4.118). Note that there are limited target cubes in $\mathscr{M}$, which are also determined by (4.116), hence the objective function aims to generate trajectory point-by-point. Thus, using the solution of (4.118), the greedy algorithm can effectively select key cubes and generate the sub-optimal trajectory for ARMS in different scenarios, respectively.

For analyzing the complexity of our algorithm, there are $V$ target cubes in total that need to be added from $\mathscr{M}$. When current location of ARMS is at the $i$th cube, it needs to compare another $|V - i|$ edges in $G$ to determine the next measuring cube. Note that every target cube contains $m$ parameters ($m = 4$ in our model), and $O(V) = O(n)$. Thus, the total operation time is $O\left(m \sum_{i=1}^{V-1} |V-i|\right) = O(n^2)$.

**Algorithm 15:** Operation of monitoring algorithm

/* ***Complete Monitoring:*** triggered between days */<br>
**for** $i = 1$ *to* $sum(Cube)$ **do**<br>
    measure the AQI value of $Cube_i$ and record;<br>
    move to the next cube;<br>
**end**<br>
generate baseline 3D AQI map $\mathbb{B}$;<br>
/* ***Selective Monitoring:*** triggered between hours */<br>
**for** $i = 1$ *to* $sum(Cube)$ **do**<br>
    calculate $PDT_{cubei}$;<br>
    **if** $PDT_{cubei} \geq PDT \mid\mid PDT_{cubei} \leq \delta$ **then**<br>
        add $Cube_i$ to $\mathscr{M}$;<br>
    **end**<br>
**end**<br>
generate min trajectory $\mathbb{D}$ of $\mathscr{M}$;<br>
**forall the** $p_i \in \mathbb{D}$ **do**<br>
    measure the AQI value of $Cube_i$ and record;<br>
**end**<br>
update the real-time AQI map $\mathbb{M}$ based on previous $\mathbb{B}$ and $\mathbb{D}$;<br>
**if** $\mathbb{M}$ *deviates* $\mathbb{B}$ *by a large* $\sigma$ **then**<br>
    enter the complete monitoring period;<br>
**end**

Algorithm 15 describes the whole process of the monitoring algorithm. Complete monitoring is triggered between days and selective monitoring is triggered between hours. When the monitoring area experiences severe environmental changes such as the gale, ARMS compares the result of map built by selective monitoring and the map built last time. If there is a large deviation $\sigma$ between them, ARMS would again trigger the complete monitoring to rebuild the baseline distribution.

### *4.5.4 Application Scenario I: Performance Analysis in Horizontal Open Space*

In this section, we implement the adaptive monitoring algorithm in a typical 2D scenario, namely the horizontal open space. We present performance analysis of GPM-NN and adaptive monitoring algorithm in this typical scenario, respectively.

#### 4.5.4.1 Scenario Description

When the 3D space has a limited range in height, ARMS needs to cover target cubes nearly in the same horizontal plane. Two distant cubes at the same height may have a low correlation, as the wind may create different concentration of pollutants in a horizontal plane. This scenario is commonly considered as a typical 2D scenario and often with a horizontal open space (e.g., a roadside park), as shown in Fig. 4.41.

**Fig. 4.41** A typical application scenario of ARMS in 2D space (a roadside park)

#### 4.5.4.2 Performance Analysis

In this subsubsection, we first compare the accuracy of GPM-NN with other existing models by the experimental result in Fig. 4.42. Then, Fig. 4.43 illustrates the influence by different numbers of neurons in the hidden layer. To study GPM-NN's performance when AQI varies, in Fig. 4.44, we show the relationship between different AQI values and corresponding estimation accuracy. In Fig. 4.45, we present the performance of our monitoring algorithm versus other selection algorithms. Finally, Fig. 4.46 shows the trade-off between system battery consumptions and estimation accuracy via different *PDT*s.

1. *Model Accuracy*: In Fig. 4.42, we compare three prediction models, our regression model GPM-NN, linear interpolation (LI) [94], and classical multi-variable linear regression (MLR) [81], respectively, versus different values of *PDT*. LI uses interpolation to estimate the AQI value of undetected cubes by other measured cubes, while MLR uses multiple parameters (e.g., wind, humidity, temperature, etc.) of measured cubes to do regression and estimation.

In the horizontal open space scenario, we can find that GPM-NN achieves the highest accuracy. In each curve, we can see that the average estimation accuracy decreases as the *PDT* value increases. As discussed in Sect. 4.5.3.2, when *PDT* has a higher threshold, target cubes in set $\mathscr{M}$ decline, i.e., the total cubes measured by ARMS become fewer. Thus, the estimation accuracy correspondingly drops. When $PDT = 0.1$, GPM-NN performs the best among three models, which proves the robust and precision of our model. Moreover, as *PDT* increases (e.g., $PDT = 0.75$), GPM-NN still maintains a high accuracy (almost 80%), while others experience a

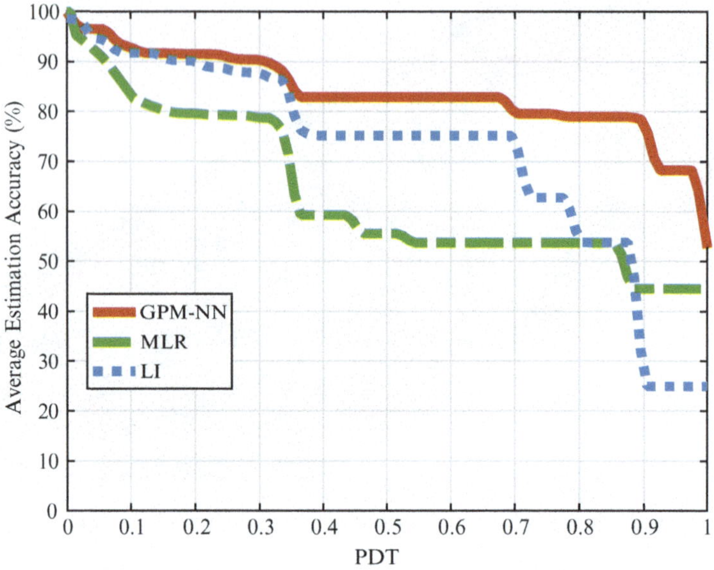


**Fig. 4.42** The comparison of estimation accuracy between GPM-NN, MLR and LI, in 2D scenario

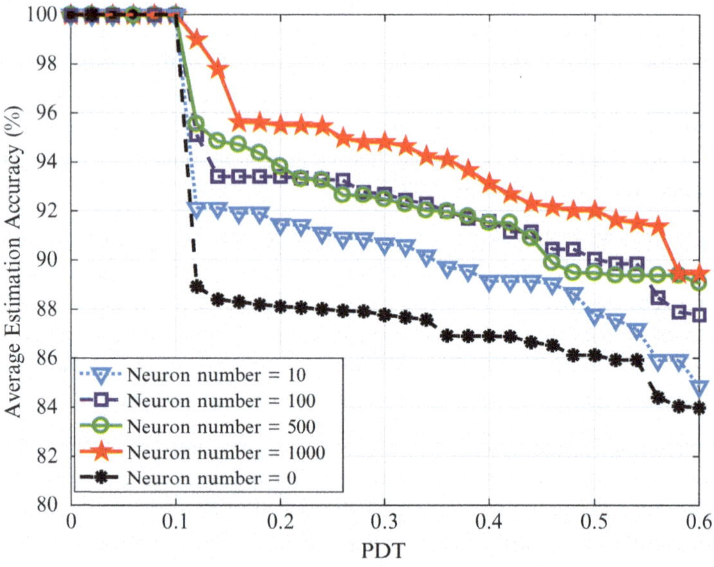


**Fig. 4.43** The impact of the number of neurons in the non-linear part, in 2D scenario

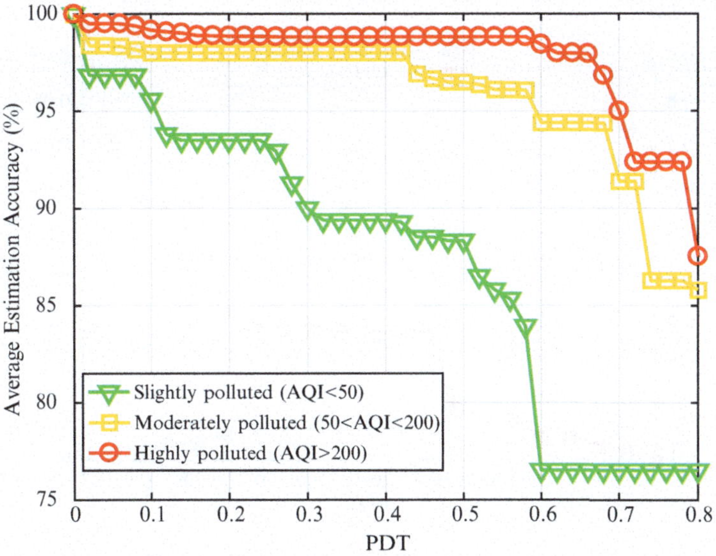


**Fig. 4.44** The performance of GPM-NN with different AQI values, in 2D scenario

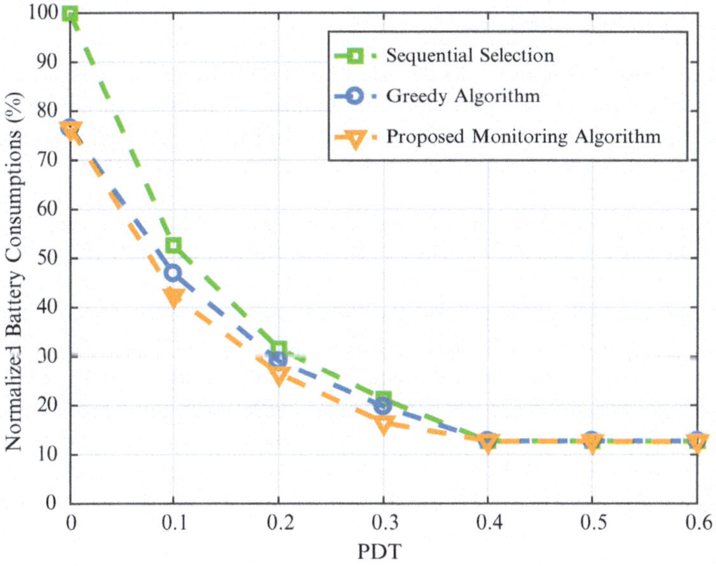


**Fig. 4.45** Comparison of the adaptive monitoring algorithm, greedy algorithm and sequential selection

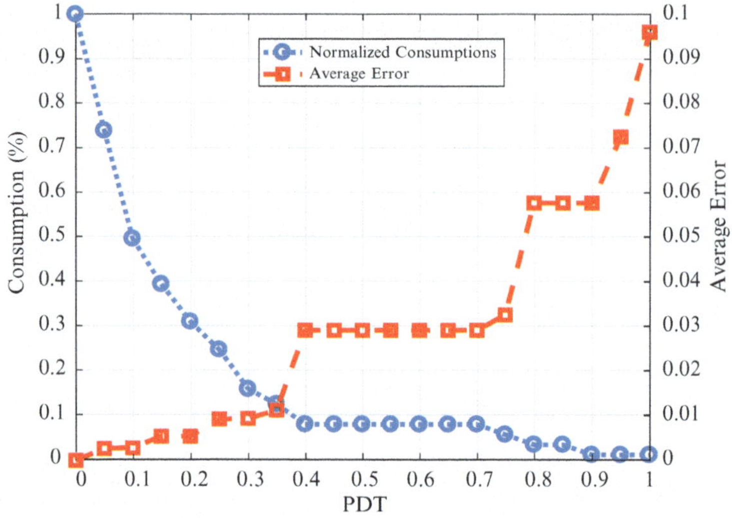


**Fig. 4.46** The trade-off between system battery consumption and estimation accuracy, in 2D scenario

rapid decrease. This implies that our model is suitable for adaptive energy saving monitoring in a fine-grained area.

2. *Effects of Neuron Numbers*: As we adopt the NN structure to introduce the non-linear part for our GPM-NN model, the number of neurons in the hidden layer can have great impacts on estimation results. In Fig. 4.43, we plot the estimation accuracy of different number of neurons in GPM-NN via *PDT*, to study their influence.

From Fig. 4.43, when $PDT < 0.1$, the monitoring contains all cubes. When the number of neurons is 0, our model is equal to the physical model in (4.102) with regression, which only contains the linear part. By comparing this curve with others, we can find out that the number of neurons $= 0$ is worse than the number of neurons $\neq 0$. By adding the non-linear part (NN structure), GPM-NN performs better with higher accuracy. Moreover, the curve with fewer number of neurons (e.g., the number of neurons $= 10$) performs worse than with more neurons (e.g., the number of neurons $= 500$). In this scenario, we can find that the number of neurons $= 1000$ can achieve the highest estimation accuracy. We ignore the situation where the number of neurons $> 1000$, as too many neurons in the hidden layer can cause overfitting.

3. *Effects of Various AQI*: In Fig. 4.44, we plot the estimation accuracy of GPM-NN with different AQI values (i.e., AQI $\leq 50$, $50 \leq$ AQI $\leq 200$, and AQI $\geq 200$ [76]), via different *PDT*s. From the curves, we can find that in 2D scenario,

GPM-NN performs the best when AQI $\geq$ 200. As 50 $\leq$ AQI $\leq$ 200, GPM-NN also maintains high accuracy, while relatively worse when AQI is low. This indicates that our model is better predicting in moderately and highly polluted days, which has great instructing significance in forecasting severe pollution as well as prevention. This characteristic is also suitable for the adaptive monitoring algorithm when AQI is high. Note that even GPM-NN performs not so good when AQI is low, it still outperforms other models.

4. *Performance of Adaptive Monitoring Algorithm*: In this part, we compare the results of the proposed monitoring algorithm for trajectory planning, versus other algorithms such as greedy algorithm and sequential selection, by plotting their battery consumptions over one measurement in Fig. 4.45. The greedy algorithm aims to select the nearest target cube in $\mathscr{M}$ to generate the trajectory [63], while sequential selection is done by selecting cubes from the bottom (or left) to the top (or right) in order [65].

In the typical horizontal open space, we plot the normalized battery consumption achieved by three algorithms in Fig. 4.45, via different *PDT*s. The normalized consumption is the cost percentage achieved by each monitoring method of one total battery charge (i.e., 15 min). As *PDT* increases, the consumption would correspondingly decrease, as the target cubes in $\mathscr{M}$ would be fewer. By comparing these three curves, we can see that the sequential selection is the most consuming method. Our monitoring algorithm performs the best and is better than the greedy algorithm, while $0.1 \leq PDT \leq 0.4$. After *PDT* reaches 0.4, the consumption of three methods becomes equal, since the target cubes in $\mathscr{M}$ now is so few that there is no difference among these algorithms. Hence, the adaptive monitoring algorithm can relatively reduce the power consumption for monitoring AQI in the 2D scenario.

5. *Trade-Off Between Consumption and Accuracy*: In Fig. 4.46, we illustrate the trade-off between the battery consumption and the estimation accuracy. To better illustrate the trade-off, we use $average\ error$ as a merit, expressed as

$$\overline{ERR} = \frac{1}{n} \sum_{i=1}^{n} \left( \frac{\hat{C}_f(i) - C_f(i)}{C_f(i)} \right)^2, \tag{4.119}$$

where $n$, $\hat{C}_f(i)$ and $C_f(i)$ are the same in (4.114). We plot the curves of system's power consumption and average estimation error versus *PDT*.

Figure 4.46 illustrates the relationship between the accuracy and the battery consumption. Intuitively, a larger *PDT* introduces a less power consumption, which proves that with a higher *PDT*, consumption declines as the number of measured cubes decreases. Moreover, when *PDT* $\geq$ 0.4, the total consumption of the whole system can be reduced by 90%. The rapid decline of consumption is also related to the high redundancy of data in the typical 2D space as the roadside park. On the other hand, the average error of ARMS increases as *PDT* becomes larger, which confirms the existence of the trade-off between power consumption and estimation accuracy. Under this circumstance, choose $PDT = 0.41$ can achieve a relatively high predicting accuracy (over 80%) while greatly reduce the battery consumption of the system.

## 4.5.5 *Application Scenario II: Performance Analysis in Vertical Enclosed Space*

In this subsection, we implement the adaptive monitoring algorithm in a typical 3D scenario, vertical enclosed space. We then present the performance analysis of the GPM-NN and the adaptive monitoring algorithms in this typical scenario, respectively.

### 4.5.5.1 Scenario Description

In the typical 3D scenario, the 3D space has target cubes in various heights. In this type of scenario, the planar area is relatively limited (e.g., the courtyard inside a high-rise building). As shown in Fig. 4.47, in such a vertical enclosed space, there is no significant difference on AQI values between two horizontally neighboring cubes, but the wind may create a discrepancy of the pollutant concentration on two cubes at different heights. Hence, the benefit of selecting more cubes vertically outweighs the cost of traversing between distant cubes at the same heights.

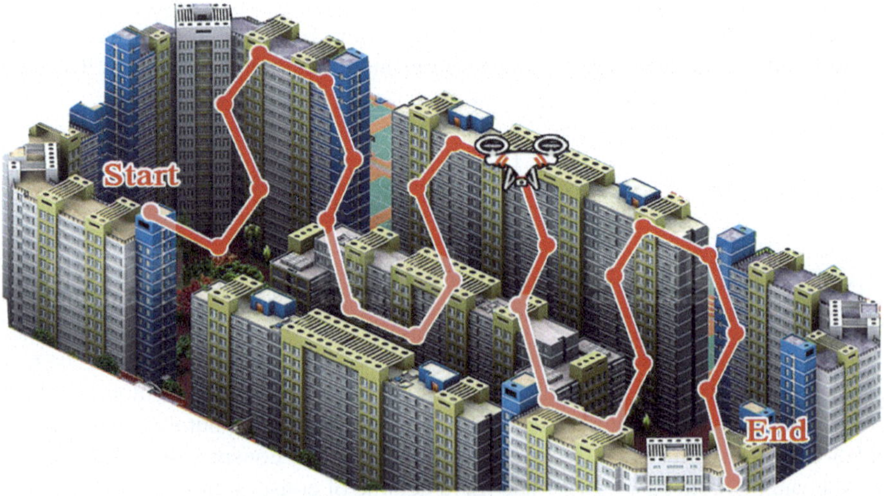


**Fig. 4.47** A typical application scenario of ARMS in 3D space (courtyard inside a high-rise building)

#### 4.5.5.2 Performance Analysis

In this subsubsection, we present performance analysis of ARMS in different aspects for typical 3D scenario.

1. *Model Accuracy*: In Fig. 4.48, we compare three prediction models. In the vertical enclosed space scenario, GPM-NN still maintains the highest accuracy among three models via different *PDT*s. Compared to the 2D scenario, LI decreases rapidly as *PDT* increases, which indicates the heterogeneity in 3D AQI distribution. Moreover, when $PDT = 0.8$, GPM-NN would experience a violent decline. This phenomenon is caused by the inherent characteristic of *PDT*. When *PDT* is high, the corresponding number of target cubes in $\mathscr{M}$ becomes so few that the predicting accuracy can significantly drop, even if only one point is unmeasured (e.g., 10 cubes with $PDT = 0.75$ and 9 cubes with $PDT = 0.8$). This result can provide the basis for choosing the suitable *PDT* value.

In conclusion, GPM-NN performs better in both 2D and 3D fine-grained scenarios, with a high estimation accuracy even if measuring cubes are few.

2. *Effects of Neuron Numbers*: In Fig. 4.49, we study the effects of the number of neurons in a typical 3D scenario. When $PDT < 0.1$, the result is the same as in the 2D scenario, that each curve performs almost the same. As *PDT* increases, the curve with the number of neurons = 0 declines most rapidly like that in Fig. 4.43. Also, the curve with fewer number of neurons (e.g., the number of neurons = 10) performs worse than with more neurons (e.g., the number of

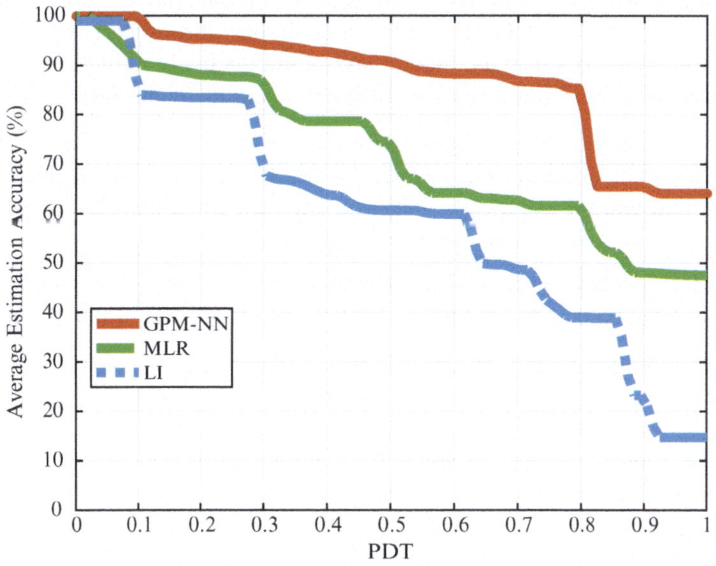


**Fig. 4.48** The comparison of estimation accuracy between GPM-NN, MLR and LI, in 3D scenario

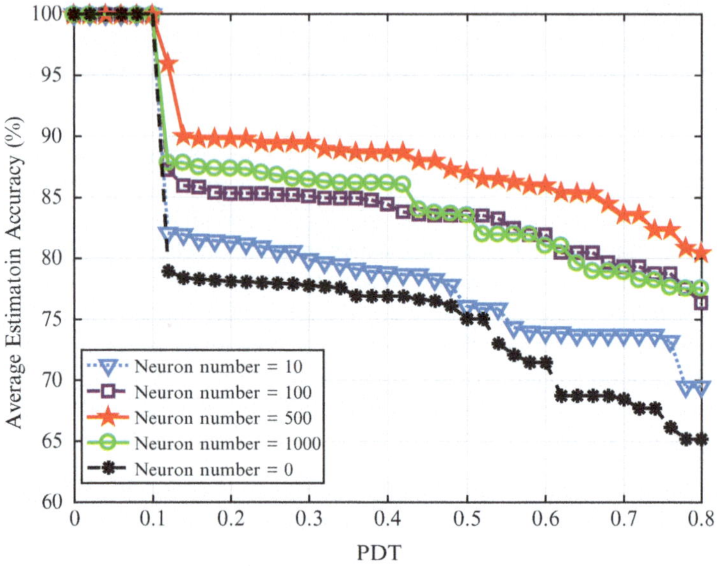


**Fig. 4.49** The impact of the number of neurons in the non-linear part, in 3D scenario

neurons = 100/1000) as well. In this scenario, we can find that the number of neurons = 500 can achieve the highest estimation accuracy, which is different from the result in the 2D scenario.

In conclusion, our GPM-NN model (with combination of linear and non-linear part) is robust and better than that with only linear part. Moreover, the number of neurons robust in the hidden layer can effectively influence the model's performance, and the optimal value is different in various scenarios.

3. *Effects of Various AQI*: In Fig. 4.50, we again plot the estimation accuracy of GPM-NN with different AQI values in the 3D scenario. From these curves, we can find that GPM-NN also performs the best when moderately and highly polluted, while relatively worse when AQI is low.

In conclusion, GPM-NN can maintain a better estimation accuracy when the AQI value is moderate and high, which is suitable for the operation of our ARMS.

4. *Performance of Adaptive Monitoring Algorithm*: In the 3D scenario as vertical enclosed space, Fig. 4.51 shows the consumption of three algorithms, our monitoring algorithm, greedy algorithm, and sequential selection, via different *PDT*s. From the figure, we can see when *PDT* is low, sequential selection consumes much more than those of our method and the greedy algorithm. This indicates that when scenario becomes 3D, the cube selection can be more complicated and a suitable selection method can highly reduce the battery consumption. Moreover, adaptive monitoring algorithm also performs the best among these

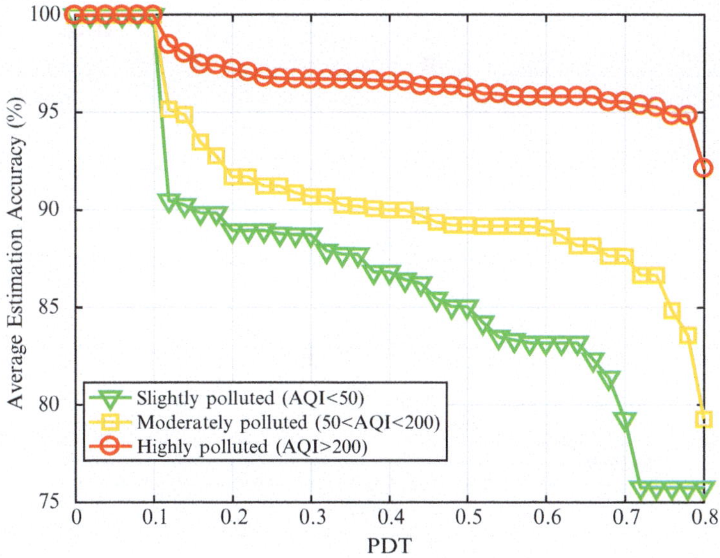


**Fig. 4.50** The performance of GPM-NN with different AQI values, in 3D scenario

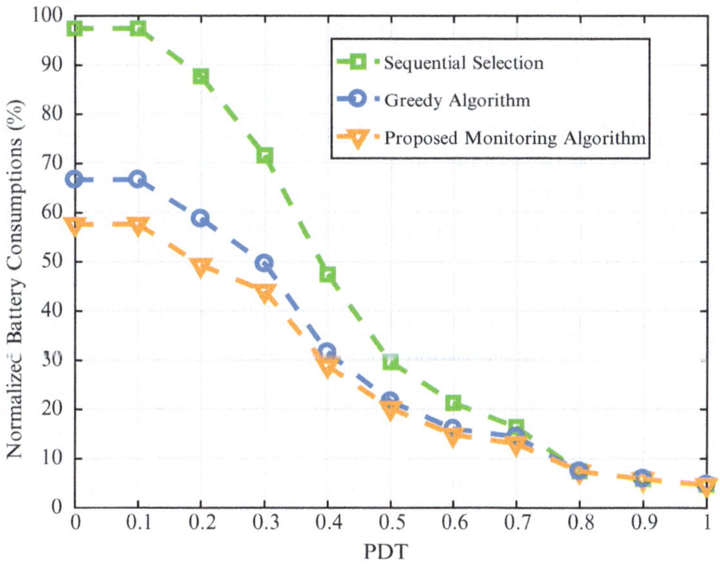


**Fig. 4.51** Comparison of the adaptive monitoring algorithm, greedy algorithm and sequential selection

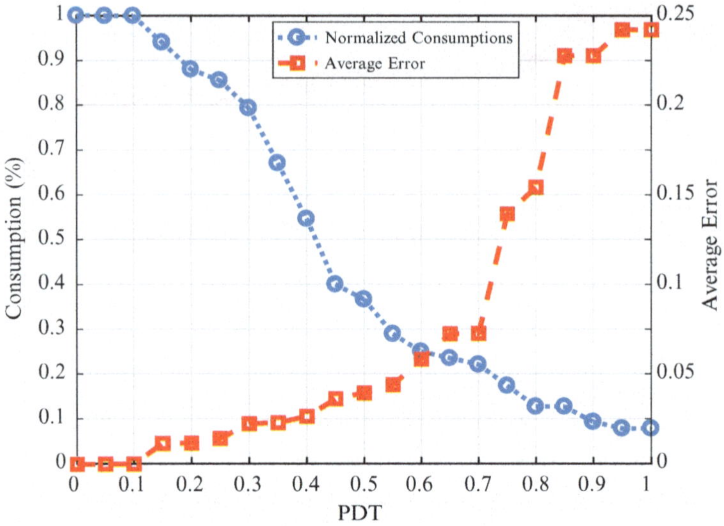


**Fig. 4.52** The trade-off between system battery consumption and estimation accuracy, in 3D scenario

three methods, and it is better than the greedy algorithm when $PDT \leq 0.8$. As *PDT* becomes higher, the normalized consumption of three algorithms is closer, and becomes equal when $PDT \geq 0.8$. Thus, the adaptive monitoring algorithm can effectively save the battery life for monitoring AQI in 3D scenario.

5. *Trade-Off Between Consumption and Accuracy*:s In Fig. 4.52, we plot the trade-off in the 3D scenario as horizontal enclosed space. This typical 3D scenario is more common in real measurement, and hence the result is more instructive. As *PDT* becomes higher, the average error grows rapidly as consumption can drop fairly. Given the average error, for example, when $\overline{ERR} = 0.04$ (average estimation accuracy is about 80%), the corresponding $PDT = 0.51$, and thus the power consumption can be reduced to as little as 37%. Hence, by choosing a suitable *PDT* value for monitoring, the measuring efforts can greatly scale down.

### *4.5.6 Summary*

In this section, we have designed a UAV sensing system, ARMS, to construct fine-grained AQI maps. A novel fine-grained AQI distribution model GPM-NN has been proposed based on NN and physical model, to help generate a real-time AQI map with data collected by ARMS. To reduce the battery consumptions of ARMS, we have proposed the adaptive monitoring algorithm to efficiently update real-time AQI maps. For the 2D and 3D scenarios, we have applied the adaptive monitoring algorithm, respectively. By using the proposed index *PDT*, the system can well balance the intrinsic trade-off between the estimation accuracy and power consumption. Experimental results have shown that GPM-NN can achieve a higher accuracy in AQI map construction than other existing models, and the number of neurons in the hidden layer of GPM-NN should also be adjusted in various scenarios to acquire better performance. Moreover, the adaptive monitoring algorithm can generate the trajectory while greatly saving the battery life of the UAV, and ARMS can well balance the trade-off between accuracy of AQI map and battery consumptions.

## References


1. L. Gupta, R. Jain, G. Vaszkun, Survey of important issues in UAV communication networks. IEEE Commun. Surv. Tutorials **18**(2), 1123–1152 (2015)
2. J. Wang, C. Jiang, Z. Han, Y. Ren, R.G. Maunder, L. Hanzo, Taking drones to the next level: cooperative distributed unmanned-aerial-vehicular networks for small and mini drones. IEEE Veh. Technol. Mag. **12**(3), 73–82 (2017)
3. N.H. Motlagh, T. Taleb, O. Arouk, Low-altitude unmanned aerial vehicles-based internet of things services: comprehensive survey and future perspectives. IEEE Internet Things J. **3**(6), 899–922 (2016)
4. S. Zhang, H. Zhang, Q. He, K. Bian, L. Song, Joint trajectory and power optimization for UAV relay networks. IEEE Commun. Lett. **22**(1), 161–164 (2018)
5. Y. Yang, Z. Zheng, K. Bian, L. Song, Z. Han, Realtime profiling of fine-grained air quality index distribution using UAV sensing. IEEE Internet Things J. **5**(1), 186–198 (2018)
6. T. Kersnovski, F. Gonzalez, K. Morton, A UAV system for autonomous target detection and gas sensing, in *Proceedings of IEEE Aerospace Conference*, Big Sky (2017)
7. B.H.Y. Alsalam, K. Morton, D. Campbell, F. Gonzalez, Autonomous UAV with vision based on-board decision making for remote sensing and precision agriculture, in *Proceedings of IEEE Aerospace Conference*, Big Sky (2017)
8. T. Zhao, D. Doll, D. Wang, Y. Chen, A new framework for UAV-based remote sensing data processing and its application in almond water stress quantification, in *Proceedings of International Conference on Unmanned Aircraft Systems*, Miami (2017)
9. S. Zhang, H. Zhang, B. Di, L. Song, Joint trajectory and power optimization for UAV sensing over cellular networks. IEEE Commun. Lett. **22**(11), 2382–2385 (2018)
10. M. Thammawichai, S.P. Baliyarasimhuni, E.C. Kerrigan, J. Sousa, Optimizing communication and computation for multi-UAV information gathering applications. IEEE Trans. Aerosp. Electron. Syst. **54**(2), 601–615 (2018)
11. A. Al-Hourani, S. Kandeepan, A. Jamalipour, Modeling air-to-ground path loss for low altitude platforms in urban environments, in *IEEE Global Communications Conference*, Austin (2014)
12. G.J. Leishman, *Principles of Helicopter Aerodynamics* (Cambridge University Press, Cambridge, 2006), pp. 159–193
13. 3GPP TR 36.777, Enhanced LTE support for aerial vehicles, Release 15, Dec. 2017
14. B.V.D. Bergh, A. Chiumento, S. Pollin, LTE in the sky: trading of propagation benefits with interference costs for aerial nodes. IEEE Commun. Mag. **54**(5), 44–50 (2016)
15. N.H. Motlagh, M. Bagaa, T. Taleb, UAV-based IoT platform: a crowd surveillance use case. IEEE Commun. Mag. **55**(2), 128–134 (2017)
16. S. Zhang, H. Zhang, B. Di, L. Song, Cellular UAV-to-X communications: design and optimization for multi-UAV networks. IEEE Trans. Wireless Commun. **18**(2), 1346–1359 (2019)

17. J. Shang, Z. Wang, Q. Huang, A robust fusion algorithm for sensor failure. IEEE Signal Process. Lett. **20**(8), 755–758 (2013)
18. J. Chen, J. Li, S. He, Y. Sun, H. Chen, Energy-efficient coverage based on probabilistic sensing model in wireless sensor networks. IEEE Commun. Lett. **14**(9), 833–835 (2010)
19. V.N. Swamy, S. Suri, P. Rigge, M. Weiner, G. Ranade, A. Sahai, B. Nikolic, Cooperative communication for high-reliability low-latency wireless control, in *Proceedings of IEEE International Conference on Communications*, London (2015)
20. S. Zhang, H. Zhang, B. Di, L. Song, Cellular cooperative unmanned aerial vehicle networks with sense-and-send protocol. IEEE Internet Things J. **6**(2), 1754–1767 (2019)
21. O.B. Sezer, E. Dogdu, A.M. Ozbayoglu, Context-aware computing, learning, and big data in internet of things: a survey. IEEE Internet Things J. **5**(1), 1–27 (2018)
22. S. Verma, Y. Kawamoto, Z. Fadlullah, H. Nishiyama, N. Kato, A survey on network methodologies for real-time analytics of massive IoT data and open research issues. IEEE Commun. Surv. Tutorials **19**(3), 1457–1477 (2017)
23. V.V. Shakhov, I. Koo, Experiment design for parameter estimation in probabilistic sensing models. IEEE Sens. J. **17**(24), 8431–8437 (2017)
24. A. Chakraborty, R.R. Rout, A. Chakrabarti, S.K. Ghosh, On network lifetime expectancy with realistic sensing and traffic generation model in wireless sensor networks. IEEE Sens. J. **13**(7), 2771–2779 (2013)
25. A. Hossain, S. Chakrabarti, P.K. Biswas, Impact of sensing model on wireless sensor network coverage. IET Wireless Sens. Syst. **2**(3), 272–281 (2012)
26. E.G. Larsson, Optimal OFDMA downlink scheduling under a control signaling cost constraint. IEEE Trans. Commun. **58**(10), 2776–2781 (2010)
27. Y. Zeng, X. Xu, R. Zhang, Trajectory design for completion time minimization in UAV-enabled multicasting. IEEE Trans. Wireless Commun. **17**(4), 2233–2246 (2018)
28. M. Chen, M. Mozaffari, W. Saad, C. Yin, M. Debbah, C.S. Hong, Caching in the sky: proactive deployment of cache-enabled unmanned aerial vehicles for optimized quality-of-experience. IEEE J. Sel. Areas Commun. **35**(5), 1046–1061 (2017)
29. M. Mozaffari, W. Saad, M. Bennis, M. Debbah, Mobile unmanned aerial vehicles (UAVs) for energy-efficient internet of things communications. IEEE Trans. Wireless Commun. **16**(11), 7574–7589 (2017)
30. Y. Zeng, R. Zhang, Energy-efficient UAV communication with trajectory optimization. IEEE Trans. Wireless Commun. **16**(6), 3747–3760 (2017)
31. H. Zhang, Y. Liao, L. Song, D2D-U: device-to-device communications in unlicensed bands for 5G system. IEEE Trans. Wireless Commun. **16**(6), 3507–3519 (2017)
32. A. Al-Hourani, S. Kandeepan, S. Lardner, Optimal LAP altitude for maximum coverage. IEEE Wireless Commun. Lett. **3**(6), 569–572 (2014)
33. D. Athukoralage, I. Guvenc, W. Saad, M. Bennis, Regret based learning for UAV assisted LTE-U/WiFi public safety networks, in *Proceedings of IEEE Global Communications Conference*, Washington (2016)
34. 3GPP TS 25.996, Spatial channel model for multiple input multiple output (MIMO) simulations, Release 6, Jun. 2018
35. M.J. Garey, D.S. Johnson, *Computers and Intractability: A Guide to the Theory of NP-Completeness*, 1st edn. (W.H. Freeman, New York, 1979), pp. 50–53
36. M. Sipser, *Introduction to the Theory of Computation*, 3rd edn. (Cengage Learning, Boston, 2012), pp. 225–277
37. M. Grant, S. Boyd, *CVX: MATLAB Software for Disciplined Convex Programming*. Version 2.1 (2016). http://cvxr.com/cvx
38. D.A. Plaisted, Some polynomial and integer divisibility problems are NP-HARD, in *Proceedings of Annual Symposium on Foundations of Computer Science*, Houston (1976), pp. 264–267
39. D. Li, X. Sun, Constrained polynomial 0–1 programming, in *Nonlinear Integer Programming*, 1st edn., chap. 11 (Springer, New York, 2006), pp. 315–348
40. S. Zhang, B. Di, L. Song, Y. Li, Sub-channel and power allocation for non-orthogonal multiple access relay networks with amplify-and-forward protocol. IEEE Trans. Wireless Commun. **16**(4), 2249–2261 (2017)

41. J. Hu, H. Zhang, L. Song, Reinforcement learning for decentralized trajectory design in cellular UAV networks with sense-and-send protocol. IEEE Internet Things J. **6**(4), 6177–6189 (2019)
42. 3GPP TR 38.901, Study on channel model for frequencies from 0.5 to 100 GHz. Release 14, Dec. 2017
43. S.O. Rice, Mathematical analysis of random noise. Bell Syst. Tech. J. **23**(3), 282–332 (1944)
44. J.I. Marcum, Table of Q functions, Tech. Rep., U.S. Air Force Project RAND Res. Memo. M-339, ASTIA Document AD 1165451, Rand Corporation, Santa Monica (1950)
45. G. Demange, D. Gale, M. Sotomayor, Multi-item auctions. J. Polit. Econ. **94**(4), 863–872 (1986)
46. R.M. Corless, G.H. Gonnet, D.E.G. Hare, D.J. Jeffrey, D.E. Knuth, On the Lambert *W* function. Adv. Comput. Math. **5**(1), 329–359 (1996)
47. R.S. Sutton, A.G. Barto, *Reinforcement Learning: An Introduction* (MIT Press, Cambridge, 1998)
48. E. Yang, D. Gu, Multiagent reinforcement learning for multi-robot systems: a survey. University of Essex Technical Report CSM (2004)
49. C.J.C.H. Watkins, Learning from delayed rewards, Ph.D. Dissertation, King's College, Cambridge (1989)
50. M. Bowling, Multiagent learning in the presence of agents with limitations, Ph.D. Dissertation, Carnegie Mellon University, Pittsburgh (2003)
51. W. Uther, M. Veloso, Adversarial reinforcement learning, Ph.D. Dissertation, Carnegie Mellon University, Pittsburgh (1997)
52. C. Claus, C. Boutilier, The dynamics of reinforcement learning in cooperative multiagent systems, in *Proceedings Conference on Artificial Intelligence/Innovative Applications of Artificial Intelligence*, Madison (1998)
53. T. Jaakkola, M.I. Jordan, S.P. Singh, On the convergence of stochastic iterative dynamic programming algorithms. Neural Comput. **6**(6), 1185–1201 (1994)
54. S. Singh, M. Kearns, Y. Mansour, Nash convergence of gradient dynamics in general-sum games, in *Proceedings of Conference on Uncertainty in Artificial Intelligence*, San Francisco (2000)
55. W. H. Organization, 7 million premature deaths annually linked to air pollution. Air Qual. Climate Change **22**(1), 53–59 (2014)
56. Q. Di, Y. Wang, A. Zanobetti, et al., Air pollution and mortality in the Medicare population. N. Engl. J. Med. **376**(26), 2513–2522 (2017)
57. Y. Li, Y. Zhu, W. Yin, Y. Liu, G. Shi, Z. Han, Prediction of high resolution spatial-temporal air pollutant map from big data sources, in *Proceedings of Big Data Computing and Communications*, Taiyuan (2015)
58. B. Zou, J.G. Wilson, F.B. Zhan, Y.N. Zeng, Air pollution exposure assessment methods utilized in epidemiological studies. J. Environ. Monit. **11**(3), 475–490 (2009)
59. Beijing MEMC, Beijing municipal environmental monitoring center (2017). http://www.bjmemc.com.cn/
60. Y. Cheng, X. Li, Z. Li, S. Jiang, Y. Li, J. Jia, X. Jiang, AirCloud: a cloud-based air-quality monitoring system for everyone, in *Proceedings of ACM Conference on Embedded Network Sensor Systems*, New York (2014)
61. D. Hasenfratz, O. Saukh, C. Walser, C. Hueglin, M. Fierz, T. Arn, J. Beutel, L. Thiele, Deriving high-resolution urban air pollution maps using mobile sensor nodes. Pervasive Mob. Comput. **16**(2), 268–285 (2015)
62. N. Nikzad, N. Verma, C. Ziftci, E. Bales, N. Quick, P. Zappi, K. Patrick, S. Dasgupta, I. Krueger, T. Rosing, W. Griswold, CitiSense: improving geospatial environmental assessment of air quality using a wireless personal exposure monitoring system, in *Proceedings of ACM Wireless Health*, San Diego (2010)

63. Y. Gao, W. Dong, K. Guo, X. Liu, Y. Chen, X. Liu, J. Bu, C. Chen, Mosaic: a low-cost mobile sensing system for urban air quality monitoring, in *Proceedings of IEEE Conference on Computer Communications Workshops*, San Francisco (2016)
64. D. Bisht, S. Tiwari, U. Dumka, A. Srivastava, P. Safai, S. Ghude, D. Chate, P. Rao, K. Ali, T. Prabhakaran, A.S. Panickar, V.K. Soni, S.D. Attri, P. Tunved, R.K. Chakrabarty, P.K. Hopke, Tethered balloon-born and ground-based measurements of black carbon and particulate profiles within the lower troposphere during the foggy period in Delhi, India. Sci. Total Environ. **573**(1), 894–905 (2016)
65. Y. Hu, G. Dai, J. Fan, Y. Wu, H. Zhang, BlueAer: a fine-grained urban PM2.5 3D monitoring system using mobile sensing, in *Proceedings of IEEE Conference on Computer Communications*, San Francisco (2016)
66. T.N. Quang, C. He, L. Morawska, L.D. Knibbs, M. Falk, Vertical particle concentration profiles around urban office buildings. Atmos. Chem. Phys. **12**(11), 5017–5030 (2012)
67. F.M. Rubinoa, L. Floridiaa, M. Tavazzania, S. Fustinonia, R. Giampiccoloa, A. Colombia, Height profile of some air quality markers in the urban atmosphere surrounding 100m tower building. Atmos. Environ. **32**(20), 3569–3580 (1998)
68. C. Borrego, H. Martins, O. Tchepel, L. Salmim, A. Monteiro, A.I. Miranda, How urban structure can affect city sustainability from an air quality perspective. Environ. Model. Softw. **21**(4), 461–467 (2006)
69. Y. Zheng, F. Liu, H. Hsieh, U-air: when urban air quality inference meets big data, in *Proceedings of ACM Conference on Knowledge Discovery and Data Mining*, Chicago (2013)
70. H.X. Xu, G. Li, S.L. Yang, X. Xu, Modeling and simulation of Haze process based on Gaussian model, in *Proceedings of International Computer Conference on Wavelet Active Media Technology and Information Processing*, Chengdu (2014)
71. M. Cameletti, R. Ignaccolo, S. Bande, Comparing spatio-temporal models for particulate matter in Piemonte. Environmetrics **22**(8), 985–996 (2011)
72. C. Zhao, M. Heeswijk, J. Karhunen, Air quality forecasting using neural networks, in *Proceedings of IEEE Symposium Series on Computational Intelligence* Athens (2016)
73. M. Cai, Y. Yin, M. Xie, Prediction of hourly air pollutant concentrations near urban arterials using artificial neural network approach. Transp. Res. Part D: Transp. Environ. **14**(1), 32–41 (2009)
74. M.W. Gardner, S.R. Dorling, Neural network modelling and prediction of hourly NOx and NO2 concentrations in urban air in London. Atmos. Environ. **33**(5), 709–719 (1999)
75. M. Dedovic, S. Avdakovic, I. Turkovic, N. Dautbasic, T. Konjic, Forecasting PM10 concentrations using neural networks and system for improving air quality, in *Proceedings of XI International Symposium on Telecommunication*, Sarajevo (2016)
76. Beijing EPB, Beijing municipal environmental protection bureau (2017). http://www.bjepb.gov.cn/
77. Plantower, Technology laser PM2.5 sensor, air quality sensor. http://www.plantower.com/en/
78. Da-Jiang Innovations Science and Technology Co., Ltd. (DJI), Phantom 3 professional. https://www.dji.com/cn/phantom-3-pro
79. Y. Yang, Z. Zheng, K. Bian, Y. Jiang, L. Song, Z. Han, Arms: a fine-grained 3D AQI realtime monitoring system by UAV, in *Proceedings of IEEE Global Communication Conference*, Singapore (2017)
80. R.V. Hogg, A.T. Craig, *Introduction to Mathematical Statistics*, 5th edn. (Prentice Hall, Upper Saddle River, 1995)
81. M. Cameletti, F. Lindgren, D. Simpson, H. Rue, Spatio-temporal modeling of particulate matter concentration through the SPDE approach. Adv. Stat. Anal. **97**(2), 109–131 (2013)
82. D.R. Middleton, Modelling air pollution transport and deposition, in *IEE Colloquium on Pollution of Land, Sea and Air: An Overview for Engineers*, London (1995)
83. J.M. Stockie, The mathematics of atmospheric dispersion modeling. SIAM Rev. **53**(2), 349–372 (2011)
84. S. Brusca, F. Famoso, R. Lanzafame, S. Mauro, A. Marino Cugno Garrano, P. Monforte, Theoretical and experimental study of Gaussian plume model in small scale system. Energy Procedia **101**(1), 58–65 (2016)

85. F. Tang, B. Mao, Z. Fadlullah, N. Kato, O. Akashi, T. Inoue, K. Mizutani, On removing routing protocol from future wireless networks: a real-time deep learning approach for intelligent traffic control. IEEE Wirelesss Mag. **25**(1), 154–160 (2018)
86. A. Al-Molegi, M. Jabreel, B. Ghaleb, STF-RNN: space time features-based recurrent neural network for predicting people next location, in *Proceedings of IEEE Symposium Series on Computational Intelligence*, Athens (2016)
87. S.M. Carroll, B.W. Dickinson, Construction of neural nets using the radon transform, in *Proceedings IEEE International Joint Conference on Neural Networks*, Washington (1989)
88. G. Cybenko, Approximation by superpositions of a sigmoidal function. Math. Control Signals Syst. **2**(4), 303–314 (1989)
89. K. Funahashi, On the approximate realization of continuous mapping by neural networks. Neural Netw. **2**(3), 183–192 (1989)
90. D.P. Bertsekas, *Nonlinear Programming* (Athena Scientific, Belmont, 1999), pp. 1–60
91. R. Penrose, A generalized inverse for matrices. Math. Proc. Camb. Philos. Soc. **51**(3), 406–413 (1955)
92. R. MacAusland, The moore-penrose inverse and least squares, in *Math 420: Advanced Topics in Linear Algebra* (2014)
93. D. Goldberg, R. Lingle, Alleles, loci, and the traveling salesman problem, in *Proceedings of International Conference on Genetic Algorithms and Their Applications*, Hillsdale (1985)
94. C. De Boor, *A Practical Guide to Splines* (Springer, New York, 1978)

The manufacturer's authorised representative in the EU is Springer
Nature Customer Service Centre GmbH, Europaplatz 3, 69115 Heidelberg,
Germany. If you have any concerns regarding our products, please
contact ProductSafety@springernature.com

Printed and bound by CPI Group (UK) Ltd, Croydon, CR0 4YY
30/09/2026
02225651-0003